The Human Advantage

The Human Advantage

A New Understanding of How Our Brain Became Remarkable

Suzana Herculano-Houzel

The MIT Press
Cambridge, Massachusetts
London, England

This book was set in Stone Sans and Stone Serif by Toppan Best-set Premedia Limited. Printed and bound in the United States of America.

Library of Congress Cataloging-in-Publication Data

Names: Herculano-Houzel, Suzana, 1972–
Title: The human advantage : a new understanding of how our brain became remarkable / Suzana Herculano-Houzel.
Description: Cambridge, MA : The MIT Press, 2016. | Includes bibliographical references and index.
Identifiers: LCCN 2015038399 | ISBN 9780262034258 (hardcover : alk. paper)
Subjects: LCSH: Brain—Physiology. | Intellect.
Classification: LCC QP398 .H524 2016 | DDC 612.8/2—dc23 LC record available at http://lccn.loc.gov/2015038399

10 9 8 7 6 5 4 3 2 1

To my parents, Selene and Darley,
who gave me wings and taught me how to fly

To Jon Kaas,
who encourages me to soar higher

Contents

Preface

Surely We Are Special—Aren't We?

Humans are awesome. Our brain is seven times too large for the size of our body, and it takes an extraordinarily long time to develop. Our cerebral cortex is the largest relative to the size of the entire brain, and its prefrontal portion is also the largest. The human brain alone costs a tremendous amount of energy—25 percent of the calories needed to run the entire body in a day. It became enormous in a very short amount of time in evolution, leaving our cousins, the great apes, behind, with their meager brains only one-third the size of ours. So the human brain is special, right?

Wrong, according to new evidence coming from my lab that you are about to discover in the chapters to come: our brain is remarkable, yes—but not special in the sense of being an exception to the rules of evolution, singled out to become amazing in its own exclusive way. And yet we seem to have the most capable brain on Earth, the one that explores other brains instead of being explored by them. If our brain is not an evolutionary outlier, then where does the human advantage lie?

The Human Advantage invites you to drop the usual bias of considering humans as extraordinary and, instead, to look at the human brain in the light of evolution and of the new evidence that suggests a different account for what makes our cognitive abilities unique: that our brain outranks that of other animals not because we are an exception in evolution, but rather because, for simple evolutionary reasons, we hold the largest number of neurons in the cerebral cortex, affordable to no other species. I will argue that the human advantage lies, first, in the fact that we are primates, and, as such, owners of a brain that is built according to very economical scaling rules that make a large number of neurons fit into a relatively small volume, compared to other mammals. Second, we are the primate species that benefited from the fact that, some 1.5 million years ago, our ancestors came up with a trick that allowed their descendants to afford a rapidly increasing and soon to be enormous number of cortical neurons, so far rivaled by none: cooking. Third, and finally, thanks to the rapid brain expansion now affordable by the extra calories, courtesy of cooking, we are the species that owns the largest number of neurons in the cerebral cortex—the part of the brain responsible for finding patterns, reasoning logically, expecting the worst and preparing for it, developing technology and passing it on through culture.

Comparing the human brain to the brain of dozens of other animal species large and small has been a most humbling experience, one that reminds me that there is no reason to suppose that we humans have been singled out in our evolutionary history or "chosen" in any way. It is my hope that this new understanding of the human brain will help us better appreciate our place on Earth as a species that, while neither special nor extraordinary (since our species conforms to the same evolutionary scaling rules that apply to other primates), is indeed remarkable in its cognitive abilities, and, thanks to its outstanding number of neurons, has the potential to change its own future—for better or for worse.

Rio de Janeiro, January 2015

Acknowledgments

This book summarizes ten years of work that were first made possible thanks to the support and generosity of Roberto Lent, who arranged for an unprecedented position in science communication to open at the Institute of Biomedical Sciences at the Federal University of Rio de Janeiro (which I filled), then invested in my crazy idea of turning brains into soup to find out what they were made of. We have since stopped collaborating, but he will always have my gratitude for setting me on my way.

Jon Kaas, Distinguished Centennial Professor at the Department of Psychology at Vanderbilt University, entered my professional life and changed its course in 2006, when we started collaborating. Ever since, we meet a couple of times a year, in scientific meetings or visits to his lab in Nashville, Tennessee, where he and his charming wife, Barbara Martin, house me (their dining room table has seen many of my papers come to life), feed me (Jon makes a mean *feijoada*), and keep my heart and mind warm with friendship and conversation. People in the field often mistake him for my former advisor (which he never was, though it would have honored me), but the truth is he's become so much more: Jon is a dear friend and a kind of scientific father to me, someone who decided to watch over me just because he could. Thank you, Jon, for everything.

I've had the good luck to run into fantastic people along the way. Bruno Mota is both a wonderful friend and great collaborator, who for more than ten years now has been there whenever things turned to math—we have great arguments over whether life is optimized (his take as a physicist) or just good enough (mine, as a biologist). Paul Manger was supposed to give me just a half brain of an elephant, but decided to trust me with dozens of brains of all sorts of creatures large and small along with his friendship. Karl Herrup, whom I've decided to consider my honorary advisor, has given me invaluable help, advice, and encouragement along the way. It's been great to have you around, guys.

The work described here was made possible by a large number of collaborators. Besides Roberto Lent, Jon Kaas, and Paul Manger, I'd like to extend my gratitude to Ken Catania, the truest and coolest biologist I know; Lea Grinberg, Wilson Jacob Filho, and their team at USP; Christine Collins and Peiyan Wong; and all the students who have participated in the studies reported here, in particular Karina Fonseca-Azevedo, Frederico Azevedo, Pedro Ribeiro, Mariana Gabi, Lissa Ventura-Antunes, Kamila Avelino-de-Souza, Kleber Neves, and Rodrigo Kazu. I've also benefited from a number of indirect collaborators, people who have helped me one way or another on my way to becoming a comparative neuroanatomist (something that I never trained to be): Georg Striedter, Patrick Hof, Rob Barton, Richard Passingham, Jack Johnson, Pasko Rakic, Chet Sherwood, Leah Krubitzer, Jim Bower, Stephen Noctor, Charles Watson, and George Paxinos. Thank you all for being there.

Scientific research in Brazil is funded entirely by federal and state funds, and I'd like to thank CNPq and FAPERJ for their financial support over the years. I'm also immensely thankful to the James McDonnell Foundation for their support since 2010, even though most of the fruits of their investment in my lab are still to appear in print.

The beautiful illustrations that permeate this book were made by Lorena Kaz, a talented young artist with whom I had the luck to spend a year in the lab thanks to a stipend from CNPq.

I thank my editor at MIT Press, Bob Prior, for taking me up on the proposition to write this book, and for his patience as new findings pushed forward the deadline a couple of times. I also thank Chris Eyer and Katherine Almeida for their editorial support, Jeffrey Lockridge for expert and careful editing, and Katie Hope for her enthusiasm in having this be a trade book.

In making this book approachable to the general readership, I had the invaluable help of my mother, a sociologist, and my daughter, then a 15-year-old, who read the first version of every single chapter, highlighter at the ready, implacably marking whatever was not readily comprehensible. Any shortcomings in the text are my fault, not theirs. (My father, on the other hand, was not of much help there, since he never had a single criticism, only more questions; those will have to wait until the next book.)

My parents never faltered in encouraging me to become what I wanted to be, a scientist, even though that was not, and still is not, a sensible career choice in Brazil (how bad of a choice is illustrated by the fact that my mother would much rather I become a musician). They taught me that it was okay to question authority, made sure I learned languages and knew how to fend for myself, then took a deep breath and gave me a plane ticket to leave the country to enter graduate school in the US at the tender age of 19, when Brazilians are typically still living at home and just starting college. My parents gave me wings and pushed me out to fly, even when it meant away from them. All I can wish for is that I make them proud.

Last but not least, my crowd at home—my children, Luiza and Lucas, and my husband, José Maldonado: Thank you guys for your patience, for putting up with my travels, with my invisibility hat (the only way to work at home, I highly recommend it!), with the often absent look on my face as I walk around the house pondering numbers in the back of my mind. Thank you for listening to the brand new discoveries-of-the-day, for cheering me on, for celebrating with me every new achievement along the way, for appreciating the effort. Your happiness is what makes it worth it.

1 Humans Rule!

So we are special—or at least that is what most books on neuroscience will state. Our brain supposedly has an impressive 100 billion neurons and ten times more glial cells, with an enlarged cerebral cortex, and it tripled in size in a mere 1.5 million years—no time at all, in evolutionary terms—whereas the brain of great apes has retained its same size, one-third of ours, for at least four times as long. Humans of the *sapiens* variety coexisted with Neanderthals and even mingled with them to some extent, but eventually only our kind prevailed. We have come to rule the world, in more ways than just dominating other animals: modern humans are the only species that can go wherever it pleases on the planet, and even beyond it.

Behind these feats is what I will call the "human advantage." As far as I know, and as presumptuous as it may seem, it is a fact that we are the only species to study itself and others and to generate knowledge that transcends what was observed firsthand; to tamper with itself, fixing imperfections with the likes of glasses, implants, and surgery and thus changing the odds of natural selection; and to modify its environment so extensively (for better or for worse), extending its habitat to improbable locations. We are the only species to use tools to create other tools and technologies that extend the range of problems it can tackle; to further its abilities by seeking harder and harder problems to solve; and to invent ways to register knowledge and to instruct later generations that go beyond teaching by direct demonstration. Even though all this may be achieved through no particular cognitive ability exclusive to our species (more on this later), we certainly take these abilities to a level of complexity and flexibility that is rivaled by none.

For decades, the human advantage seemed to lie in a number of features that appeared to make our brain an oddity, an exception to the rules.

Gorillas are about two to three times as large as we are, but they have a brain that is only one-third the mass of ours, which makes the human brain seven times too large for our body mass. This enlarged human brain also costs far more energy to operate per day than seems reasonable: a whole one-third of the energy needed to run the entire rest of the body, muscles and all, even though it represents a meager 2 percent of our body mass. The rules that apply to other animals don't apply to us. It seems only fitting, then, considering how our achievements set us apart from all other living beings, that our extraordinary cognitive abilities must require an extraordinary brain.

Given what it can achieve, the human brain most certainly is remarkable. But is it really out of the ordinary? That is the central question explored in *The Human Advantage*. Is our brain indeed made of 100 billion neurons and ten times more glial cells, as many respected authors have long affirmed? (No, in fact, it's not.) Is it really seven times too large for the size of our body? (It is, but only if humans are compared to great apes, which turn out to be the odd ones out, instead of us.) Does it really use an extraordinary amount of energy? (Not for its number of neurons, no.) And if it turns out that the human brain is not extraordinary, then how can it still achieve such remarkable feats?

And how did we humans, and no other species, come to have such remarkable cognitive abilities—what happened in evolution that led our species to rule over all others? How come it was humans, and not great apes, who gained so much larger a brain in so short a time? Has evolution been all about a progression of life-forms that culminated in humans, its crowning achievement?

Humans on Top: Evolution as Progress

Not surprisingly, the history of how the human brain came to be considered special is intertwined with the history of evolution itself—and, for a long time, both were stories of too many interpretations based on too few facts.

Life changes over geological time, and has been changing ever since it first appeared, some 3.7 billion years ago. This is a fact, for it does not depend on interpretation, just as it is equally a fact that no humanlike beings older than 4 million years are found in the fossil record: we are a very recent "invention." These facts of changes in life over time, which go by the name of "evolution," were only recognized less than 200 years ago. Ever since, the very concept of evolution has been evolving, from that of progression toward perfection to that of simple change over time, the modern view, as this chapter will make clear. Failing to recognize evolution, however, never stopped humans from studying at least some of its facts: the wonderful diversity of life-forms that it creates.

In the face of diversity, our brain automatically creates categories to which even the unruliest forms of diversity are assigned. Just as writing implements are categorized as "pens" or "pencils" and wheeled vehicles as "cars," "trucks," or "bicycles," life-forms visible to the naked eye have been categorized, at least since the time of Aristotle, some 2,300 years ago, as "plants" or "animals." But Aristotle went further, and conceived of a "Great Chain of Being"—a *scala naturae* or nature's ladder—that arranged all things natural in a fixed hierarchical ladder of categories in descending order, from the Prime Mover at the top down to minerals at the bottom, with animals somewhere in between, arranged "by degree of perfection of their souls."[1] On nature's ladder, as it came to be accepted over the centuries, humans came second only to God.

Until the concept of evolution as change over time appeared, that hierarchy was fixed: life-forms in all their categories had always been and would always be the same—and naturalists framed their thoughts and observations on the diversity of life according to that unchanging ladder of nature. In the eighteenth and nineteenth centuries, however, the uncovering of growing numbers of particular fossils in geological layers of a certain age led inexorably to the new concept of the mutability over time of the full panoply of living beings—and evolution came to be conceptualized for the next generations by Charles Darwin in 1859. In the light of evolution, nature's ladder gained a time axis, and for many it became an evolutionary ladder which organisms supposedly ascended as they evolved, over time, from simple to complex. Instead of being fixed, that great ladder of nature was now seen as telescoping over time (figure 1.1), extending ever upward, toward humans. It was therefore fitting that humans appeared only recently in the fossil record.

Thus reasoned German neurologist Ludwig Edinger, by many considered the father of comparative neuroanatomy. At the end of the nineteenth century, Edinger viewed brain evolution (in line with Darwin)

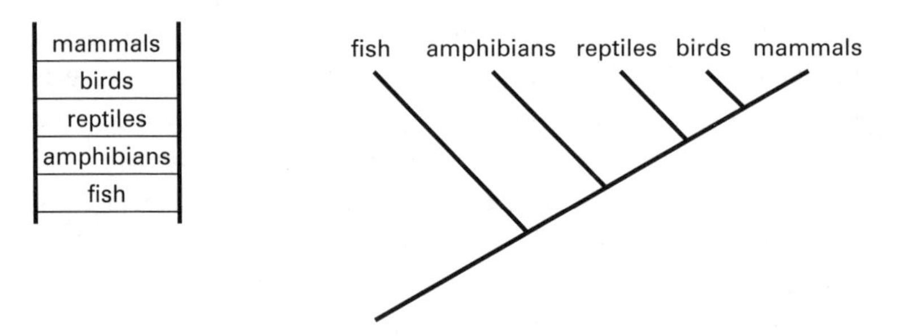

Figure 1.1
Simplified version of nature's ladder for vertebrate animals (*left*), and the same scale now stretched over evolutionary time as it became understood that life evolved, that is, changed over time (not drawn to scale). The merging lines (*right*) indicate that modern birds and mammals (aligned at the top) shared a common ancestor, and their common ancestor shared a common ancestor with modern reptiles, and their common ancestor shared a common ancestor with modern amphibians, and so on, back to the first life-form on the planet. This particular "genealogical tree" of vertebrates is wrong, though; see figure 1.4.

as progressive and also unilinear (in line with the telescoping version of Aristotle's ladder, unfolding over evolutionary time): from fish to amphibians, then to reptiles, birds, and mammals—culminating with the human brain, naturally, in an ascent from "lower" to "higher" intelligence, according to what supposedly was the order in which the different vertebrate groups appeared on Earth. In the process of climbing up the ladder, Edinger reasoned, the brains of existing vertebrates retained structures of those which preceded them. Therefore, in the face of progressive evolution, the comparison of the brain anatomy of existing species should reveal the origin of the more recent structures from older ones. The supposed evidence of "past lives" in modern brain structures resonated with the law of recapitulation that had been formulated by German embryologist Ernst Haeckel in 1886 in the aphorism "Ontogeny recapitulates phylogeny" (that is, development recapitulates evolution): Haeckel claimed that the development of more recent ("advanced") species passed through successive stages represented by adult forms of older (more "primitive") species. Edinger extended to the adult brains of different species what Haeckel thought he saw in their embryos.

And so it was that, in the beginning of the twentieth century, and in line with the idea of progressive evolution from fish to amphibians,

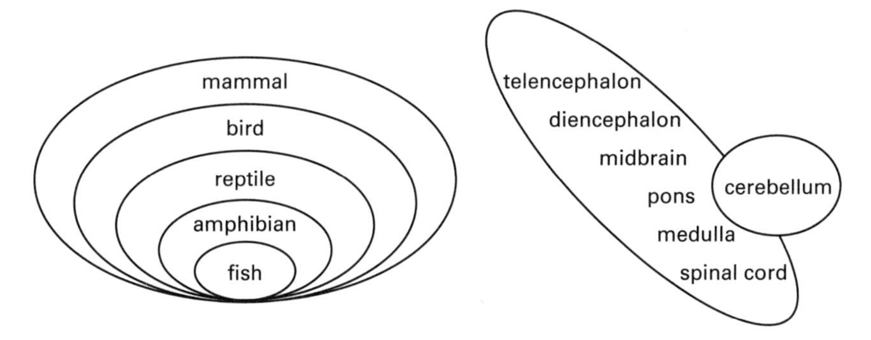

Figure 1.2
In Edinger's view, just as mammals would have evolved by progressing past a bird-like stage, and birds in turn would have progressed past a reptile-like stage, the brain of each vertebrate group would have acquired new structures on top of those in preexisting species (*left*). The resultant layering of structures was reminiscent of the sequence of structures along the vertebrate brain and spinal cord (*right*), from top (telencephalon) to bottom (spinal cord).

to reptiles, to birds, to mammals, and to humans in particular through gradual increases in complexity and size, Edinger suggested that each new vertebrate group in evolution acquired a more advanced cerebral subdivision, much as the Earth's geological layers formed over time, one on top of the other (figure 1.2). The layering of those subdivisions was reminiscent of the main divisions of the human central nervous system (spinal cord, medulla, pons, cerebellum, diencephalon, mesencephalon, and telencephalon), recognizable in all vertebrates. Fittingly, the telencephalon—the top layer, and therefore supposedly the most recent—is the one that differs the most in size across species, and stands out very clearly in the human brain, where it accounts for almost 85 percent of all brain mass (figure 1.3).

Edinger proposed, in 1908, that the preeminence of the telencephalon in mammals, and particularly in humans among all mammals, was a sign of the human evolutionary status as the "highest" among animals. He held that the telencephalon itself had also evolved progressively, through the addition of layers: an ancestral part of the telencephalon (the striatum) controlled instinctive behavior, and had been followed by the addition of a newer brain (the pallium or cortex), which controlled learned and intelligent behavior—and that part was the most developed

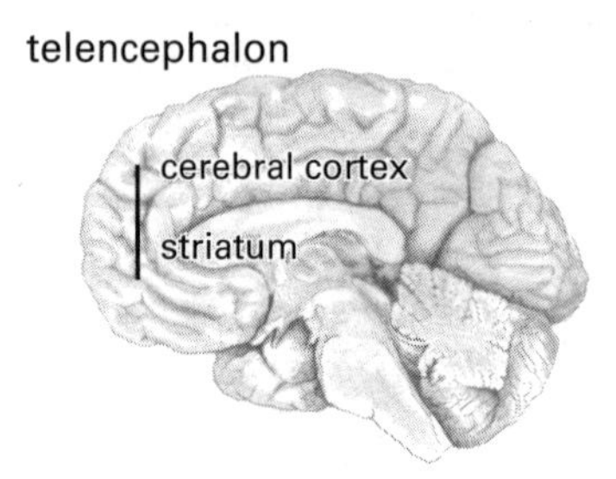

Figure 1.3
Although the human brain has the same subdivisions as all vertebrate brains, the human telencephalon (cerebral cortex + striatum) is several times larger than all other brain structures together.

in humans.[2] As it went, the primordial telencephalon of fishes had a small cortex and a larger striatum to which another level of striatum and cortex was added in reptiles. Birds would have evolved a hypertrophied striatum, but not any further pallial regions; in contrast, mammals were thought to have evolved the latest and greatest achievement, on top of the primitive cortices—the neocortex. This view was to become dominant in neuroscience, codified in an important comparative neuroanatomy text in 1936.[3]

The (mistaken) idea that the neocortex was a recent mammalian invention layered on top of older structures gained currency enough to reach the twenty-first century when neuroanatomist Paul MacLean proposed in 1964 his view of a "triune brain," which consisted of a reptilian complex (from the medulla to the basal ganglia), to which had been added a "paleomammalian complex" (the limbic system), and later a "neomammalian complex"—the neocortex.[4] The intuitive (but incorrect) equation of evolution with progress, along with the alluring notion of a primitive, reptilian brain supposedly incapable of anything as complex as a mammalian neocortex could achieve, attracted much attention from the media when it was featured, in 1977, in Carl Sagan's popular book *The Dragons of Eden*.[5]

The triune brain is only a fantasy, however. As more and more fossils of sauropsids (the proper name for dinosaurs) were uncovered, some of them feathered, it became clear that modern lizards, crocodiles, and birds are close cousins, all now considered reptiles (birds included), whereas modern

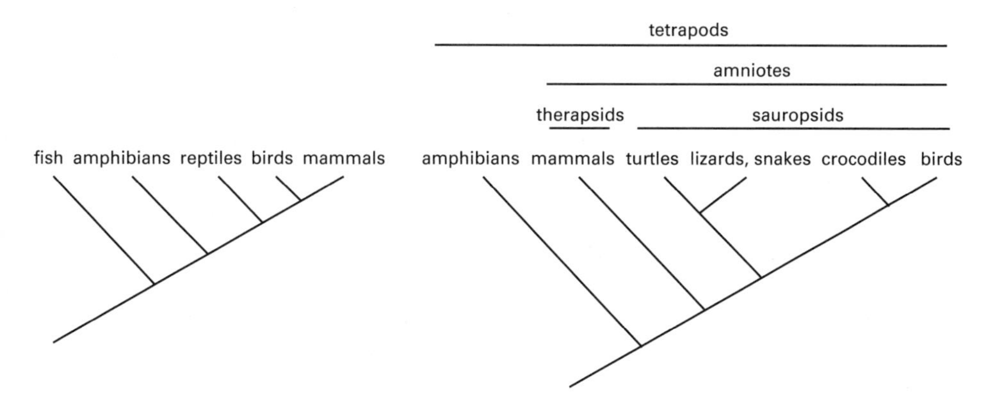

Figure 1.4
Modern rendition of fact-based evolutionary relationships among tetrapod vertebrates (*right*), in contrast to the initial view based on the stretching of nature's ladder over evolutionary time (*left*). Mammals (the modern therapsids) and reptiles (the modern sauropsids, which include birds) are sister groups. Mammals could therefore never have descended from reptiles.

mammals arose separately, and very early on, from a sister group at the very beginnings of amniote life[6] (figure 1.4). Mammals were therefore never reptiles or birds back in evolutionary time; the brain of mammals is at least as old as the brain of birds and other reptiles, if not older—it just evolved along a different evolutionary path. Indeed, modern neuroanatomical studies showed that the "striatum" of birds* has the same organization and function as the cortex of mammals: they are simply two different layouts for a structure that works in very much the same way.[7] And if mammals did not descend from reptile-like beings, they could not have a brain that was built by adding layers on top of that reptilian-like brain. Comparing the mammalian to the reptile brain and presuming one to have layered new structures on top of the other's is just as preposterous as looking at two living human cousins and expecting one of them to have given birth to the other. Still, the notion of an ancestral "reptilian" brain has been hard to shake off, and many good neuroscientists, without

*To avoid confusion with the proper striatum, or basal ganglia, the bird "striatum" was for a while called the dorsal ventricular ridge, but is now called the "pallium" (meaning "mantle"), similar in function to parts of the mammalian cerebral cortex.

proper training in evolutionary biology, have until recently compared mammal brains to reptile or bird brains as if they were looking into the evolutionary past.

Edinger's evolutionary ascending scale of brains also fails in the face of an evolutionary fact: species do not always "progress" into more complex life-forms as they evolve.[8] Certainly, the most complex beings found at any point in evolutionary time have become more and more complex as life evolved; but much simpler, unicellular life-forms still dominate the Earth's biomass, and examples abound of species that have become smaller and less complex over evolutionary time, such as microbats and intracellular parasites. Evolution is not progress, but simply change over time.

The notion that evolution retraces the steps of development of earlier species, creating new ones by adding extra levels to preexisting developmental programs—another pillar of Edinger's view—was itself refuted by evidence from the study of embryonic development in the twentieth century.[9] Modern evolutionary developmental biology understands that differences among adult animal species arise because of evolutionary modifications in their developmental program—that is, that phylogeny occurs through changes in ontogeny, the exact opposite of what Haeckel had advocated. When the developmental program changes, new life-forms arise, neither more nor less "advanced": just different.

Still, by building on the "evolutionary" version of nature's ladder, Edinger established the basis of a nomenclature that was used for an entire century to define the cerebral subdivisions of all vertebrates—and one that, through MacLean's popular triune brain model, influences popular concepts of brain evolution to this day. Even more important, Edinger started a school of thought that equated "brain evolution" with "progress," and that was to provide the backdrop to one of the most pervasive accounts for what gives humans their cognitive advantage over other animals: the notion that the human brain—or the cerebral cortex, or the prefrontal cortex, or the amount of time it takes to mature, or the amount of energy it consumes—is "larger than it should be."

The Larger-Than-It-Should-Be Human Brain

Larger animals, be they vertebrates or invertebrates, tend to have larger brains. This relationship was recognized and formulated as early as 1762,

Figure 1.5
Larger animals usually have larger brains: a rat brain, at 2 grams (about 1/14 ounce), is much smaller than a capybara brain (75 grams; about 3 ounces), which is smaller than a gorilla brain (about 500 grams or 1.1 pounds), which in turn is much smaller than an elephant brain (4,000–5,000 grams or 9–11 pounds). However, the relative size of the brain, that is, the fraction of body mass that it occupies, is smaller in larger animals, which becomes evident when these animals are drawn as if they had the same body size (*lower row*).

when Swiss naturalist Albrecht von Haller proposed what became known as "Haller's rule": that larger animal species have larger brains—although, as larger brains appear, these become smaller relative to body size[10] (figure 1.5).

The relatively smaller brain of larger animals is an example of allometric growth, or "allometry"—as opposed to isometric growth, which would happen if larger animals had proportionately larger brains as a constant fraction of body mass. The field of allometry, the study of how body shape and the proportion of body parts vary depending on body size, can be traced in spirit back to Galileo Galilei in the seventeenth century. Galileo recognized that the bodies of larger animals (and their bones, in particular) could not be proportionately (isometrically) scaled-up versions of the bodies of smaller animals, or they would simply collapse under their own weight—much like what we intuitively acknowledge that should happen to Salvador Dali's elephants on spindly legs.

The term "allometry," however, was only coined in 1891 by German physician Otto Snell,[11] and later elaborated upon by Julian Huxley, a British biologist who gave mathematical treatment to the relationships between body size and shape depicted famously in the beautiful drawings of D'Arcy Wentworth Thompson in his 1917 book *On Growth and Form*. Huxley observed that allometric relationships always took the form of power laws, where one parameter varies in proportion to a second parameter raised to a certain exponent, rather than multiplied by a constant, as in linear relationships (figure 1.6). That the mass of body parts always relates to body mass according to power laws of the form $Y = bX^a$ is today well understood: the power law is the only function that defines a *scale-invariant* relationship between X (for example, body mass) and Y (mass of body part, e.g., brain). In the case of mammals, scale invariance describes how mammalian bodies vary in mass by eight orders of magnitude (that is, by a factor of about 100 million) while maintaining a recognizable common overall architecture.

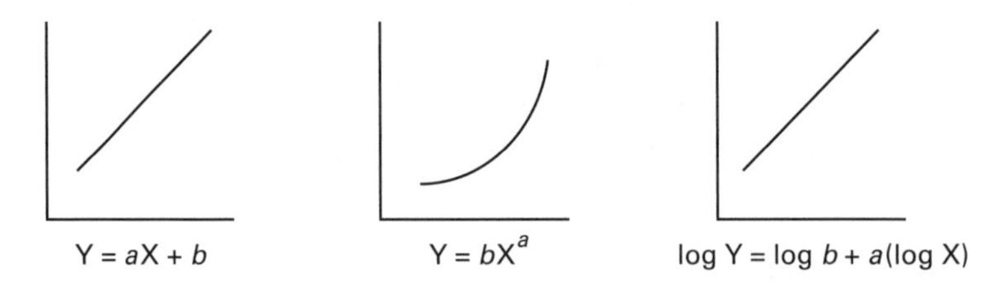

Figure 1.6
Linear function plotted on a linear scale (*left*), power function (where the allometric exponent $a > 1$) plotted on a linear scale (*center*), and relationship between log-transformed values plotted on a linear scale (*right*), which turns the power function into a linear function, much easier to calculate in the days before digital computers. In allometric functions, X is body mass, and Y is typically the mass, volume, or surface area of a body part.

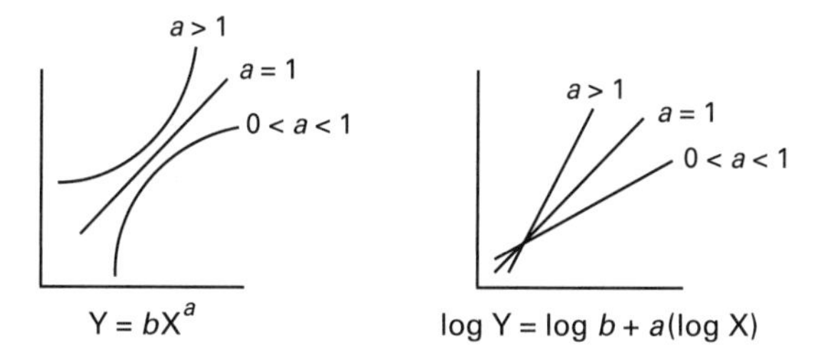

Figure 1.7
Power laws when plotted directly on a linear scale (*left*) and when plotted as the log-transformed values of X and Y on a linear scale (*right*), depending on the allometric exponent a of the function. When $a > 1$, Y increases faster than X (body mass), as with bone mass; when $a = 1$, Y increases proportionately to X, as with blood volume; and when $0 < a < 1$, Y still increases with X, but more slowly, as with brain mass.

In practice, this means that a mammal is always easily recognized as a mammal, regardless of its size. This fact implies that there must be general quantitative biological rules, or scale-invariant laws, relating the various body parts that remain valid over the entire range of brain sizes—and the application of these laws in nature, as bodies are built, is evidenced in the allometric relationships described by Huxley.

The allometry of a body part, such as the brain, describes how that part scales (varies in size with respect to surface, mass, or volume) as the body scales in mass,* and can be described by the allometric exponent of the relationship to body mass. As illustrated in figure 1.7, allometric exponents of 1.0 turn power laws into linear functions, and therefore imply that that part of the body scales isometrically (proportionately) with the body as a whole, as happens with the volume of circulating blood, which is always a fixed proportion of body volume. Exponents larger than 1 indicate that a body part grows faster than the body as a whole, which is the case of bones, as predicted by Galileo. And allometric exponents smaller than 1 but still larger than 0 indicate that a given body part grows with the body, but at a slower pace, lagging behind it such that its proportionate mass is smaller in larger animals. This is the case of the brain.

*Mass (in g) and volume (in cc, or cm^3) are mostly interchangeable measures of three-dimensional size, proportional to one another by a factor of 1.036, given that animal bodies are mostly composed of water.

Figure 1.8
Plotted line is the allometric function brain mass = $b \times$ body mass^a, calculated for the data points shown, where each point represents one mammalian species. The line illustrates the predicted brain mass for an animal of any given body mass; knowing that body mass allows one to predict, by simply applying the formula, how large the brain of that animal should be. Occasionally, however, a species (filled circle) is found to have a much smaller brain, and another (filled square) is found to have a much larger brain, than it "should." Of course, whether a species with a larger-than-expected brain has a brain too large for its body or a body too small for its brain is a whole other issue.

The recognition by North American neuropathologist Gerhardt von Bonin in 1937 that the relationship between brain mass and body mass across many different species could be described by power laws with a given allometric exponent allowed a new concept to emerge: that of brain enlargement relative to the size *expected* for a given body mass. The reason is that the existence of an allometric relationship for brain mass, defining a power law that describes how large mammalian brains are inside bodies of a certain mass, allowed the prediction of the size of the brain that a generic mammalian species "should" have for its body mass. That prediction could then be compared to the *actual* brain mass of each species (figure 1.8).

Such comparisons between what the mass of a brain structure "should" have been and what it actually was for a given body mass were made in 1969 by Heinz Stephan and Orlando Andy, working in what had been Ludwig Edinger's laboratory in Frankfurt, Germany. Stephan and Andy assumed (wrongly, but that much would only be established some forty years later) that primates, and most if not all modern placental mammals, had their evolutionary origin in insectivore-like animals like moles and shrews, living animals that, because of their diminutive size, were at the time considered to be similar to what the ancestral mammals must once have been like. In good Edingerian tradition, Stephan and Andy calculated what they called "progression indices": a measure of how much modern species would have distanced themselves from the "primitive" state of the ancestral brain, supposedly similar to a modern insectivore's brain. They first calculated the brain volume expected for a "primitive" animal from the allometric relationship between brain volume (or the volume of each particular structure) and body mass that applied to species they considered to be basal insectivores, that is, the closest and therefore most similar to the ancestral mammal. Next, they calculated their progression indices: "how many times larger a given brain structure of a certain species is than the corresponding structure in a typical basal insectivore of the same body weight."[12] The larger a brain structure was, compared to how large it should be in a basal insectivore of a similar body mass, the more "progressive" it was inferred to be.

As Edinger himself would have predicted, Stephan and Andy found that the neocortex shows the strongest "progression" in primates, in an "ascending primate scale," whereas the olfactory bulb was the only structure found to regress, that is, to become smaller than predicted in a body of that size, compared to the "primitive state." Progression was particularly high, and actually highest, in the *human* neocortex: it was the one that deviated the most from the expected volume for the corresponding body mass. It was thus that Stephan and Andy dutifully concluded that the size of the neocortex "represents the best cerebral criterion presently available for the classification of a given species in a scale of increasing evolutionary stages." Their findings seemed to prove Edinger right: "progressive evolution" was about the enlargement of the neocortex relative to the body and to other brain structures, and that enlargement "culminated in mammals, and especially in primates"[13]—and was maximal in humans. There it was: humans had an overly large neocortex compared to what a "primitive" mammal would have.

The Encephalization Quotient

The concept of a progressive index came to make history, however, under another name: the "encephalization quotient," presented in the seminal

1973 book *Evolution of the Brain and Intelligence* by the North American paleontologist Harry Jerison.

Jerison became interested in allometry as described by Huxley in his *Problems of Relative Growth*[14] and applied the concept to how the volume of the brain changes as body mass increases. Jerison built on the 1897 work of the Dutch paleoanthropologist Eugène Dubois (the discoverer of *Homo erectus*) on "indices of cephalization," the ratio between brain and body mass, which in turn was developed from Otto Snell's 1891 analysis of the brain to body relationship across species. Whereas Stephan and Andy were calculating allometric relationships for basal insectivores and extending them to make predictions about primate brains, Jerison opted instead to calculate the allometric relationship for brain and body mass to predict brain size in a wide sample of different mammalian species all mixed together.[15]

The predicted brain mass could then be compared to the *actual* brain mass of each species, as illustrated in figure 1.8. In Jerison's view, growing bodies necessarily required a growing brain mass to deal with their sensory information and to control their movements. Therefore, if the brain mass of a species was greater than expected for its body mass (given the relationship that applies to other species), that species had "extra" brain mass beyond what was strictly necessary to deal with its body. To use Jerison's term, it was more "encephalized"—and therefore more intelligent.

By dividing the actual, observed brain mass of the most diverse mammalian species by the predicted brain mass calculated for their body mass, Jerison now had a single adimensional number for each species, the encephalization quotient, that described how much larger a species' brain was relative to how large it "should" be. For him, that quotient had the explicit purpose of serving as an indicator of intelligence both in human evolution (he calculated the encephalization quotient of extinct human ancestors by using cranial capacity to estimate brain volume), and across primate and nonprimate species. By his calculations, the modern human encephalization quotient was around 7.5, which meant that the volume of the human brain was about seven-and-a-half times larger than the volume of the brain of a generic mammal of our body mass was calculated to be.[16] In a distant second place came assorted primate species, with meager encephalization quotients of around 2. Human evolution, he proposed well in line with Edinger's concept of progressive evolution, had been all about an advancement of encephalization quotients that culminated in man. That settled the question. What made us humans, what distinguished the human brain from the brains of other species and explained our remarkable and unsurpassed cognitive abilities, even if ours was not the largest brain on the map (elephant and some whale brains are larger), was the fact that our brain was *much larger than it should be* for our body mass. We were outliers, the exception to the rule that larger animals have larger brains by only so much. We were special, indeed.

The Problem with Encephalization

Over the following decades, the encephalization quotient became widely accepted as a standard for comparing species, with the assumption that it served as a better proxy for cognitive capacity than absolute brain size.[17] In retrospect, however, there had been ample evidence to question that simple explanation of "larger than it should be." Granted, any "excess brain mass" should be available for functions other than those related to bodily demands: reasoning, recognizing patterns, planning for the future. But allometric relationships are calculated by fitting an equation to the data points, which necessarily leaves some data points above and others below the fitted curve, as in figure 1.8. Therefore, if there are species that are encephalized (because they fall above the curve, with brain volumes larger than expected), by mathematical necessity, there are other species that fall below the curve and are therefore "*under*encephalized," having *less* brain mass than required to operate the body—a highly unlikely proposition given that those species are very much alive and well.

Even more problematically, however, the circular assumption that the encephalization quotient expressed intelligence or any quantitative indication of cognitive capacity (because it is maximal in humans and therefore must be related to intelligence) was not founded on tried-and-true correlation with actual measurements of cognitive capacity. Moreover, there were incongruities to the presumed direct correlation between encephalization quotient and cognitive capabilities. Second to humans alone, the fairly small capuchin monkey sports the next highest encephalization quotient among primates—around 2—and therefore would be expected to outsmart great apes,[18] whose puny encephalization quotients fall well below 1, which in theory makes great apes some of those species that are missing part of the brain mass necessary to operate the body. Yet great apes unquestionably have more complex and flexible behaviors than capuchin and other small, more encephalized monkeys and marmosets.[19] Jerison's encephalization quotient worked fine to set humans apart from other species, but it left other species in a muddle much in need of sorting out.

Additionally, the concept of encephalization assumed that all brains were made the same way, that is, that a certain amount of brain mass always contained a similar number of neurons across species. But "extra brain mass" by, say, a factor of 2 should amount to larger absolute brain mass, and therefore a larger number of "extra neurons," in a larger brain than in a smaller one. In practice, then, the same encephalization quotient should mean a larger cognitive advantage in a larger brain than in a smaller one. But this was hard to sort out, first, because comparing cognitive capabilities across nonhuman species is quite difficult and, second, because no one really knew at the time how many neurons different brains had. But what did it matter, anyway? Criticizing encephalization seemed like splitting hairs; if the human species was by far the most encephalized, wasn't that enough?

The encephalization quotient remained the most used parameter to compare species, in the hope that it indeed reflected something akin to intelligence, for three decades. It was the only metric, as opposed to absolute or relative brain size, by which the human species obviously stood out in comparison to all others.[20] Its use was embraced by North American biologist Lori Marino, a passionate defender of cetaceans' right to personhood, who found that dolphins have encephalization quotients above 3, smaller than humans but well above those of all other primate species.[21]

But then, in the first decade of the twenty-first century, the meaning of the encephalization quotient started to be questioned in the literature,[22] and much-needed systematic comparisons started to be made of general cognitive capabilities across nonhuman primates[23] and of specific self-control abilities across mammals and birds.[24] Not surprisingly, the common finding of these comparisons was that simple absolute brain size was a much better correlate of cognitive capabilities than the encephalization quotient. It was back to square one. If the human brain is not the largest, then how can it be the most capable of them all?

Special in More Ways than One

Never mind that the encephalization quotient turned out not to be the best correlate of cognitive capabilities among nonhuman animals. For many specialists, and especially for most nonspecialists, the largest-brain-for-its-body entry remained in the top of the list of "features that make the human brain special." After all, if larger animals have larger brains and gorillas are up to three times our size, then their brains should be larger than ours—and yet, the human brain is about three times as large as the gorilla brain. Jerison had made it clear: the human brain is extraordinarily large for the body that houses it.

Not special enough? Consider the amount of energy that a human brain consumes: about 500 kilocalories per day. That is about 25 percent of the energy that the entire human body requires—a hugely disproportionate amount of energy, considering that the human brain only amounts to 2 percent of the entire body mass. In comparison, a mouse's brain (which accounts for about 1 percent of its body mass) costs only 8 percent of the energy that runs its entire body.[25] The human brain is extraordinarily expensive.

The list of distinctive features of the human brain continued to grow as more and more scientists in different disciplines joined the quest for what underlies our remarkable cognitive abilities. Large spindle-shaped cells called "Von Economo neurons" were discovered initially in the anterior cingulate cortex of the human brain, an area involved in cognitive and emotional control of behavior, and in great apes, but not in other primates nor other mammals.[26] Von Economo neurons were quickly hailed as a hallmark of the human brain and even tentatively associated with consciousness.[27] But, as investigations were extended to other species, it became clear that other primates also had them,[28] and indeed these neurons are found in brains both larger and smaller than the human brain.[29]

The distinctions of the human brain are not limited to its neurons. Nancy Ann Oberheim and colleagues found in 2009 that human astrocytes, glial cells that are crucial to synaptic transmission of information across neurons, are much larger than astrocytes in the brains of other primates and of rodents, which means that our astrocytes should support many more synapses than astrocytes of these other species.[30] It could still be the case that the distinction is simply a matter of brain size; Oberheim and colleagues examined no brains larger than the human brain, and elephant astrocytes, for example, may turn out to be even larger than ours. Nevertheless, human astrocytes do seem "better" than rodent astrocytes: when human glial progenitor cells were transplanted into the developing mouse brain, which resulted in human astrocytes populating the brain of the adult mice, the recipient animals were turned into faster learners.[31]

At a higher level of organization, where neurons assemble into networks, the human brain is also often found to differ from the runner-up and frequent measuring stick to which the human brain is compared: the chimpanzee brain. For instance, the "neuropil," the fraction of the cerebral cortex taken up by synaptic connections, is larger in the human than in the chimpanzee brain[32]—although that could be simply due to the larger mass of the human brain, rather than its being special. Some scientists have also reported human-specific brain regions and networks not present in the monkey brain,[33] although other scientists disagree.[34] On the other hand, brains as different as those of humans, monkeys, cats, and even pigeons have been found to be organized as similar small-world networks, with similar hubs that concentrate and distribute signals, such as the prefrontal cortex and hippocampus.[35]

More recently, the focus has shifted to studies of our genetic makeup, which have been generating an ever-growing list of genes that are human specific, that is, "different" in humans compared to chimpanzees, our closest cousins, in the hope that those genes will turn out to hold the secret to our uniqueness. The list includes genes that control brain size,[36] synapse formation,[37] speech and language development,[38] and cell metabolism,[39] as well as those controlling other human-specific morphological features such as the shape of the human wrist and thumb.[40]

Yet, in 2003, when I became interested in brain diversity and evolution and, in particular, in how the human brain compares to others, I realized that, despite the wealth of complex data on genes, functional areas, and particular cell types, we still did not understand the basics: what brains are made of—how many neurons, how many glial cells in each brain structure. We had no idea how the number of neurons, the information-processing units of the brain, compared between human and other brains. Genetic and cellular particularities aside, could it be that the simplest reason for our remarkable cognitive capabilities, matched by none, was a remarkable number of neurons?

2 Brain Soup

Such was the state of affairs when I became interested in what brains were made of and how the human brain in particular compared to others: it didn't. Because, with that long and still growing list of oddities, we were special.

It baffled me, having trained in biology, that so many of my fellow neuroscientists would accept such a statement of human singularity without question, and even endorse and propagate it. Ever since Darwin, we have strived to understand those rules that apply to all living beings, those constraints shared by all living matter. We have come to understand that, behind the diversity of all mammals, there is a basic, heritable, genetic makeup that restricts it: an apple can only fall so far from the tree. But if those evolutionary constraints also apply to humans, then how could the human brain, and it alone, be at the same time so similar to others in the evolutionary rules that it obeys, and yet so different—to the point of endowing us with the ability to ponder our own material and metaphysical origins, while other animals stick to their own turf quite literally?

By the first decade of the twenty-first century, extraordinary features of relative size, energy cost, and a particular genetic makeup of the human brain seemed all the more necessary to account for our remarkable cognitive abilities because what little work had been done in the century before to compare actual brain matter of humans with that of other species seemed, again and again, to place humans nowhere special. For technical reasons, comparing brain tissue of different species was for decades limited to estimating densities of neurons in sections of that tissue, that is, to measuring how many neurons were visible under a microscope in a section of certain dimensions. A handful of scientists, Donald Tower and Herbert Haug foremost among them, examined the density of neurons in the cerebral cortex of assorted mammalian species—and did not find humans to stand out in any way. In terms of anatomy, our brain matter seemed to be made not very differently from that of other animals. If that was the case, then our superiority had to lie in something else that was out of the ordinary, like the larger relative amount of brain matter in the body, or its larger relative use of energy, as if our brain were an overclocked computer.

Or perhaps the reason human brain matter seemed so ordinary was that humans were simply being compared to the wrong species, like apples to oranges. In his pioneering work, Donald Tower,[1] and later Herbert Haug,[2] would freely compare mice, rabbits, cats, cows, whales, monkeys, and humans to one another with no regard to the possibility that if there are obvious external differences across rodents, carnivores, artiodactyls, cetaceans, and primates, then there might also be at least some internal differences as well, possibly extending to their brains. Harry Jerison would do the same. Such comparisons were fully in line with the underlying assumption at the time that all mammalian brains were made the same, with a shared relationship between brain size and number of neurons—and that the human brain was, in effect, no different from that of any other mammal. Although never stated explicitly, the expectation throughout the twentieth century, evident in the usual practice of comparing brains indiscriminately across species, was that all mammalian brains large and small were scaled-up or scaled-down variations of the same basic blueprint.

But what if they weren't?

Are All Brains Made the Same?

Let's put encephalization and energy cost aside and step back to reconsider the basics for a moment. A little bit of critical thinking tells us that all brains can't be made the same way, with the same and universal relationship between brain size and number of neurons. All brains are, indeed, made of neurons, the smallest computational units that process information and pass it on in the large-scale networks that they construct in the brain. Information is received through synapses, and there are an estimated 10,000 to 100,000 synapses per neuron. Information conveyed by all of these synapses converges on the cell body of the neuron, which then processes it and passes the result on to the next neuron. Although having more synapses certainly must contribute to increasing the flexibility and complexity of information processing in a neuron, it is reasonable to suppose that, ultimately, the computational capacity of the brain of a given species should be much more limited by its number of neurons than by its number of synapses.[3]

If all mammalian brains were made the same way, then two brains of a similar size should be made of similar numbers of neurons, distributed in a similar fashion across their structures. And if numbers of neurons are limiting to cognitive capability, then cows and chimpanzees, both owners of brains of about 400 grams (14 ounces), should be comparable in cognitive prowess. It has been surprisingly difficult to devise ways to compare the cognitive capabilities of different species, however. Cognitive tests must be ecologically meaningful to the species tested: they must honor not only their particular interests but also their body conformations (do they have hooves, claws, fingers—or wings?) and yet translate *across* them.* That said, we are familiar enough with these two particular species to suspect that chimpanzees have much richer, more flexible, and complex behavioral repertoires than cows. Unless, of course, cows lead very rich internal mental lives and are so smart that they choose not to let us see that they are practicing deep thinking while grazing. A highly unlikely possibility, considering that cows let themselves be raised for meat in peaceful herds, whereas chimpanzees devise mischievous ways to fool their caretakers in zoos. Clearly, two brains of similar size *don't* necessarily have similar cognitive abilities.

Still, let's play along because it's about to get worse. If all mammalian brains were built the same way, varying in size as larger or smaller versions of a basic plan scaled according to the same rules, then larger brains should always have more neurons than smaller brains, and, since neurons are the basic computational units in the brain, larger brains should for this reason also be more cognitively capable than smaller brains. But here comes that most vexing of all problems related to brain size: the human brain is not the largest of them all. It is actually nowhere close to being the largest,

*A very promising stab at cross-species comparisons of cognitive capabilities was made in 2014 by a large team of scientists around the globe led by Evan MacLean. More on this later.

which is why its largest-of-all encephalization quotient was welcomed as a much awaited solution to this paradox. At least thirteen species have brains as large as or larger than our average 1.5 kilograms (3.3 pounds)—and the largest known brain, at 9 kilograms (about 20 pounds), a full six times as heavy as ours, belongs to the sperm whale. The honor does not go to cetaceans alone: the brain of the African elephant, at 5 kilograms (11 pounds), weighs more than three times as much as ours.

Sperm whale	9,000 g	Grey whale	4,300 g
Fin whale	6,930 g	False killer whale	3,650 g
Blue whale	6,800 g	Pilot whale	2,700 g
Killer whale	5,600 g	Bowhead whale	2,700 g
African elephant	5,000 g	Minke whale	2,300 g
Humpback whale	4,700 g	Bottlenose dolphin	1,500 g
Asian elephant	4,400 g	Humans	1,300 g

These incongruities indicate that not all brains are made the same way, nor scale the same way—and made me suspect that maybe we didn't actually know what different brains, much less the human brain, were made of. Although the literature held many studies on the volume and surface area of the brain of different species,[4] and various papers on densities of neurons in the cerebral cortex,[5] estimates of numbers of neurons were scant. In particular, I could find no original source to the much-repeated "100 billion neurons in the human brain." At best, that would be an order-of-magnitude approximation. In 1988, Robert Williams and Karl Herrup had estimated "about 85 billion neurons" in the human brain, from partial estimates for the cerebral cortex and cerebellum, but they have since been repeatedly miscited as having established a much rounder average number of 100 billion neurons in the human brain. In 2003, I consulted several senior neuroscientists, most of whom believed in the 100 billion figure—but not one could refer me to the original citation. I later ran into Eric Kandel himself, whose textbook *Principles of Neural Science*,[6] a veritable bible in the field, proffered that number, along with the complement "and 10–50 times more glial cells." When I asked Eric where he got those numbers, he blamed it on his coauthor Tom Jessel, who had been responsible for the chapter in which they appeared, but I was never able to ask Jessel himself. Apparently, a partial approximation had been taken for an order-of-magnitude estimate, which in turn had been taken for the real thing. It was 2004, and no one really knew how many neurons could be found on average in the human brain.

It now seems that the ten-times-more-glial-cells-than-neurons is an even bigger myth than the one-hundred-billion-neurons figure. According to the myth, if there were indeed one hundred billion neurons in the brain as a whole, there would have to be one trillion glial cells as well. It was known, by 2004, that there are fewer than 2 glial cells for every neuron in the gray matter of the human cerebral cortex, and far fewer than 0.1 in the cerebellum.[7] That meant there would still have to be close to one trillion glial cells in the striatum, diencephalon, and brainstem, which together weigh less than 200 grams. If that were true, then there should be about 5 million glial cells per milligram of tissue in these areas, whereas, in all areas examined by 2004, densities of glial cells were found to be never higher than 100 thousand per milligram of tissue, a number fifty times too small. The popular estimates of 100 billion neurons and ten times more glial cells in the human brain seemed to be the result of a game of "telephone" among scientists, where acceptance of a handful of early order-of-magnitude estimates and glia/neuron ratios for the real thing spiraled out of control and, attractive round numbers that they were, made it solidly into the imagination of neuroscientists and the public alike.

One of the reasons for the absence of factual whole-brain estimates of numbers of cells was that the only reliable way available then to count cells was through stereology, using for example the "optical fractionator," the most trusted stereological method for counting cells. The optical fractionator consists of placing virtual three-dimensional probes throughout thin slices of tissue that single out a known small fraction of the overall tissue volume; counting the number of cells found within the probes; and then extrapolating to the total number of cells in the entire tissue volume. This is fine for tissues with a relatively homogeneous distribution of cells, or at least with only small variations in the density of cells across portions of the tissue. The highly heterogeneous distribution of neurons across different structures of the brain, however, makes using stereology impractical for determining numbers of cells in whole brains. Neuronal densities vary by factors of up to 1,000 across brainstem structures and the cerebellum. Even within a single structure, like the cerebellum, different layers have neurons packed in widely varying densities. Dealing with that in a stereological study would require parceling the brain into hundreds of structures of comparable densities, then placing a very high number of counting probes across them. It would be prohibitively expensive even for a well-equipped lab. And it was even more so for myself, with no lab and no funding.

How to Count Cells with No Lab?

When I first got involved in the business of counting cells, I was no neuroanatomist, and stereology was nowhere near my area of expertise. In 2003, I didn't even have a lab, much less funding. I had been hired just the previous year to work at the Federal University of Rio de Janeiro as an assistant professor specializing in science communication, which is what I had done for the previous three years at a science museum in Rio, after a fairly unorthodox scientific upbringing. After training in virology as a biology undergraduate at the Federal University in Rio and a brief stint as a would-be geneticist at Case Western Reserve University (CWRU) in Cleveland, I was initiated in neuroscience by a friend from graduate school and trained in the development of the peripheral nervous system in Story Landis's lab for two years at CWRU, where I discovered systems neuroscience while taking graduate courses. I then moved to Germany, where I got my Ph.D. in visual neurophysiology while working in Wolf Singer's lab at the Max Planck Institute for Brain Research in Frankfurt. After deciding not to pursue postdoctoral studies in neuronal oscillatory responses, which had been my research subject there, I got a job as a visiting scientist at the Museum of Life back home in Rio, to accompany my then husband, who had just become a postdoctoral fellow in Roberto Lent's lab at the Federal University there. For the first three years back in Rio, I designed hands-on activities for the children who visited the museum, created a website, and wrote my first book on neuroscience for the general public, which landed me a job at my alma mater. My primary duty was to train young scientists in science communication, but the hiring committee made it clear that I would also be free to pursue research activities again if I so chose.

I quickly took them up on their offer. My curiosity had been kindled while digging through the literature to research a topic that had come up while I was working at the museum. I had thought that the best way to start on a job dealing directly with the public was to have an idea of their thoughts on the brain, so, in 1999, I had conducted a survey of more than 2,000 respondents called "Do You Know Your Brain?"[8] It was a printed quiz with ninety-five short statements like "There is no consciousness without a brain," "Drugs cause dependence because they act on the brain," and "The mind is a product of spirits, not the brain," to which respondents had to answer, "Yes," "No," or "I don't know." One of the statements was "We only use 10 percent of the brain," to which 60 percent of college-educated Cariocas (Rio natives) answered, "Yes." I was surprised: I had seen the catchy phrase used in popular science magazines and even in publicity pieces, but I had no idea that it had such a strong hold on the public's imagination—especially because it's a myth. We use 100 percent of the brain, all the time, as we learn and progress and achieve great things and even as we sleep; we just use it differently over time.

But what if only 10 percent—or even fewer—of the brain's cells were neurons? It said so in *Principles of Neural Science*—that we had 100 billion neurons and 10 to 50 times more glial cells—and it had become such an accepted "fact" that neuroscientists were allowed to start their review papers with generic phrases to that effect without citing references. It was the neuroscientists' equivalent to stating that genes were made of DNA: it had become a universally known "fact." If we had ten times more glial cells than neurons, then neurons would be roughly 10 percent of all brain cells, and if neurons were the cells that really matter in terms of cognition (at which glial experts would rightly roll their eyes), then it could indeed be argued that we only used 10 percent of our brain cells.

But what if it just wasn't true?

Digging through the literature for the original studies on how many cells brains were made of, the more I read, the more I realized that what I was looking for simply didn't exist. For all the many ideas and even the seeming consensus, we actually did not understand the first thing about how many cells constituted a brain, much less how the human brain compared to others.

Stereology had not given the answer, would probably not give one, and I would not be able to afford it anyway. But what if I didn't have to? Among the studies I had read on attempts in the 1970s to determine numbers of cells in the brain were a handful of papers proposing to extract and measure total DNA from the brain and to divide that amount by the average DNA content per cell nucleus to arrive at the estimated total number of cells in the brain.[9] It would work, yes—but all I could think of, having years before seen my undergraduate advisor harvest cell nuclei from cultures of cells, was saying "No, don't measure DNA, count nuclei!" I had an idea. I was going to turn brains into soup.

The Answer Is in the Soup

If the major obstacle to counting cells in the brain was the heterogeneity of their distribution in the tissue, I could literally dissolve that heterogeneity in detergent. That is, if I could somehow dissolve only the cell membranes in the tissue, but not the nuclear membrane in each cell, then I could turn the brain into a soup of free-floating cell nuclei and easily count them by sampling just a few aliquots (tiny amounts) of the suspension, made homogeneous by agitation. As long as every cell in the brain had one and only one nucleus, if I knew how many cell nuclei there were in a brain, it followed that I would also know how many cells it had. If only I had a lab to work in.

Roberto Lent, head of the Federal University's Department of Anatomy at the time and one of the people responsible for hiring me, certainly had a lab—and, ironically, he was just in the process of finishing a neuroscience textbook called in English *One Hundred Billion Neurons*. When I asked him if he knew where that number came from, as I expected, he did not. "What if I told you I knew how to get a proper estimate?" Roberto was willing not only to listen to my strange idea of turning brains into soup to count cells but also to offer me bench space and supplies in his lab, even though, if my idea worked, he might have to change the title of his book (which he later did, though barely; because the first edition was already a success in sales in Brazil by the time we had our results, later editions simply gained a question mark: *One Hundred Billion Neurons?*).

There was a kitchen blender forgotten on a high shelf in his lab, and human brains would be available from the Pathology Department over at the University Hospital. I would start with mice and rats, of course, and most likely not use a kitchen blender; still, in those early days, every time I walked into his lab, I would catch a glimpse of that blender and ask myself, "Am I one day really going to throw a human brain in a blender and dissolve the very essence of what once was a person?" The image of the harsh, sudden, and complete annihilation of a human brain bothered me. But I finally convinced myself that, ultimately, dissolving a brain would not be so different from cutting a brain into tens of thousands of minuscule pieces, which is what anatomists routinely have to do to be able to visualize single brain cells under the microscope. The difference was that, instead of *cutting* the brain into tiny pieces, I would *dissolve* it into even smaller pieces: the cell nuclei themselves. Besides, the final protocol did not involve anything violent like throwing pieces of brain into a blender. It was much more of a kitchen recipe, complete with slicing, dicing, and crushing bits of tissue into mush.

My initial attempts did involve a small blender, though. Collecting cell nuclei is standard practice in biochemistry, so I started testing borrowed protocols that involved deep-freezing rat brains in liquid nitrogen to crack the cell membranes and then tossing the rock-hard tissue in a hand-held kitchen blender. The result should have been easy enough to foresee: bits and pieces of frozen rat brain hurled all over the lab, of course. Turning brains into soup to estimate numbers of brain cells would only work if I could collect every last nucleus from each brain. Having bits of frozen brain randomly fly out of the soup, even if only into the blender lid, was obviously not acceptable.

Dissolving lightly fixed tissue in a hand-held glass homogenizer (figure 2.1) proved much more promising. Fixation in paraformaldehyde cross-links protein molecules in the tissue and renders it hard and mechanically stable. Because the nuclear membrane is highly protein rich, it becomes very well fixed and resistant to mechanical stress.

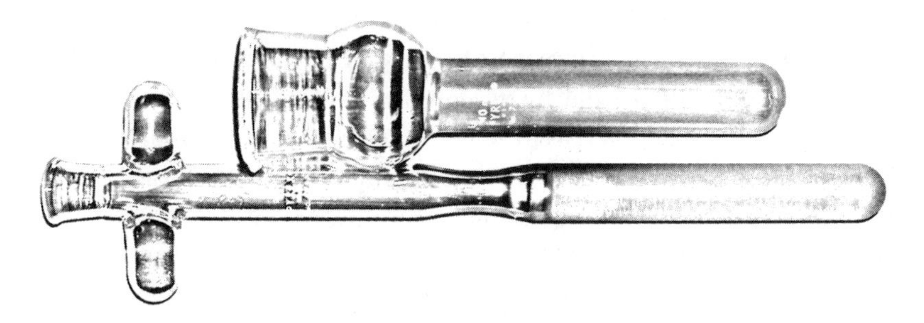

Figure 2.1
Glass tissue grinder, used like a cylindrical mortar and pestle to homogenize brain tissue.

The first attempts with fresh, unfixed tissue showed what nuclei looked like if destroyed: they turned into wads of free DNA, tainted blue under the microscope because of the DNA-binding fluorescent dye DAPI that I used to make the nuclei visible. Preparations made from tissue that had been fixed for just a few hours had a larger proportion of intact nuclei, but many were still broken.

I struck gold with hard-fixed tissue that had sat in fixative for about two weeks. That was the idea that made it all possible: if even brief fixation protected at least some of the nuclei from destruction in the homogenizer, then a very thorough, long fixation, to the point where the tissue became nearly rock-hard, might ensure that every single nucleus survived the process—which was, after all, the whole point of the new method. It finally worked: when every attempt with a hard-fixed brain yielded very similar numbers, I knew I had a new, efficient, and effective way to count cells.

After some fiddling with details such as how best to collect the nuclei and to transfer them to graduated tubes with negligible loss, I had a stable protocol. It consisted of first dissecting the hard-fixed brain into smaller, anatomically and functionally meaningful areas, such as the cerebral cortex as a whole, the cerebellum, the olfactory bulbs, and "the rest" (for now). Once weighed, each brain part was sliced and diced by hand into smaller portions that would facilitate the dissociation process: the sliding back and forth in Triton X-100 detergent between the glass walls of the homogenizer, which dissolves the cell membranes but keeps the freed cell nuclei intact. Twenty minutes of up-and-down twisting movements of the piston into the tube later, I had a cloudy, speck-free suspension of nuclei sloshing in the homogenizer: the brain had been completely turned into soup. The next step was to make sure that all nuclei to be counted had been collected, which I did by washing the piston into the tube a few times, using a pipette to collect the nuclei directly from the bottom of the tube, then washing the walls of the tube again and again and collecting all the washes into the one final volume of free nuclei, to which I added the blue fluorescent DNA dye DAPI and saline solution as required to bring up the volume to a round number that could be read accurately on the graduated cylinder. Staining one extra final wash with DAPI to look at under the microscope assured me that there were no nuclei left behind in the homogenizer: all the nuclei from all the cells in the tissue were now stained blue and collected in a known volume, ready to be counted. All it took now was agitating the suspension to distribute the nuclei evenly in the liquid before taking a few aliquots to count, which would be representative of the whole.

Counting the free cell nuclei under the fluorescence microscope required no special training: they were the roundish things that were far too large to be bacteria or mitochondria. Using a hemocytometer, a special slide-and-cover-slip chamber with a precise volume of 4 nanoliters (4 millionths of a milliliter) above each of 25 squares etched onto the glass, I could easily count how many nuclei there were in 100 nanoliters of the suspension, and proportionally calculate how many there were in the known total volume of free nuclei. The entire process at the microscope only required ten minutes to count nuclei in four aliquots of the suspension. Provided that the suspension had been made homogeneous by gentle agitation right before collecting each aliquot, the coefficient of variation across four aliquots was usually less than 0.10—that is, the standard deviation of the four counts represented no more than 10 percent of the average of the four counts. With that little variation, the estimate of the total number of cells in the tissue was as reliable as estimates obtained with stereology.

With total cell count in hand, I took advantage of the existence of an antibody that binds specifically to a protein expressed in essentially all neuronal cell nuclei, and only in them: neuronal nuclear protein or "NeuN," discovered in 1992,[10] when its function was still unknown.* The important thing about NeuN was that, even though a few particular neuronal types here and there do not produce it, for my whole-brain cell-counting purposes, I could safely consider NeuN expression to be a good marker for all neurons, and for neurons alone—and the presence of NeuN could be detected in the free cell nuclei by adding antibodies labeled red to the

*The NeuN protein was later shown to bind specific RNA sequences and regulate mRNA (messenger RNA) splicing in the cell nucleus. It is now also called Fox-3, after it was found, bizarrely enough, to be a homologue of the Fox gene ("Fox" standing for "Feminizing locus On X") involved in sex determination in the nematode worm *Caenorhabditis elegans*. Since whatever determines sex can hardly be expected to be directly involved in neuronal function, this is probably one of many examples of genes that get co-opted for different functions in different stages of development or evolution.

suspensions. It only required a small volume of the suspension to react against the red dye–labeled anti-NeuN antibody, and, after a few hours, I could take the nuclei back to the microscope to determine what percentage of all nuclei (stained blue) had belonged to neurons (now stained red). Counting 500 nuclei (which took around fifteen minutes at the microscope) was enough to determine the percentage of neurons with a certainty of 0.2 percent. Applying the percentage of neurons to the total number of cells in the structure of origin yielded an estimate of the total number of neurons in it. By subtraction, I had the total number of other cells, presumably mostly glial cells, in the tissue. Summing the results for the various brain structures—and I started with entire brains subdivided simply into cerebral cortex, cerebellum, and rest of brain—yielded, for the first time, direct estimates of the total numbers of neurons and other cells in the whole rat brain. And the entire process took less than a day.

I was thrilled. I knew something that nobody else in the world knew at that point—how many neurons an entire rat brain was made of.

The next question, of course, was whether the numbers were good—and good in the field of quantitative neuroanatomy meant comparable to those obtained with stereology. No comparison would be possible for estimates of cells in structures not amenable to stereology; that, after all, was the whole point of creating a new method, to go where stereology could not. Luckily for us, however, there were a few stereological estimates available in the literature for the rat cerebral cortex and cerebellum—and our estimates matched them.

In 2004, Karl Herrup, then at Case Western Reserve University, came to a symposium that Roberto and I organized in Caxambu, Brazil, to speak on the importance of understanding numbers of cells in the brain. Karl had long been interested in counting cells in the cerebellum, his favorite part of the brain, but had abandoned the idea for lack of an appropriate method (the cerebellum is particularly refractory to stereology because of the extremely high density of tiny neurons in the granular layer, often too close to each other to be counted separately with any accuracy). He had been the thesis advisor to a dear friend of mine, which meant I had become a fixture in his lab at CWRU years before and came to consider him my honorary advisor. When I explained to Karl in Caxambu the gist of our method for counting brain cells, he smiled: "I had thought of something like that years ago; I wanted to use flow cytometry to count cells freed from the tissue. But I never pushed forward with it. You scooped me, and I'm glad you did!" Upon learning that the paper describing the method had yet to be submitted, he immediately offered to receive it as editor at the *Journal of Neuroscience*. And so, in 2005, after a very reasonable round of peer review, Roberto and I got our first paper published in one of the most respected journals in the field.[11]

Although turning brains into soup to count the total numbers of cells in them gave results comparable to stereology (where there were results to be compared) and yielded data that were more and more consistent as we extended our analysis to more species, our method met with considerable resistance from some specialists, particularly those who saw their favorite theories begin to fall apart in the face of the new numbers. Reviewers and critics alike wanted to see the same proof: a side-by-side comparison of the new method with the well-established stereology. I was not set up to do stereology, so for many years I would not be able to run the verification myself. In the end, it was better that I wasn't the one to prove us right. That was done independently, in 2014, by Christopher von Bartheld, from the University of Reno, and by Daniel Miller and my soon-to-be collaborator Jon Kaas, from Vanderbilt University in Nashville.[12] Ever since Jon and I started collaborating in 2006, his lab had been adapting our method to obtain even faster, automated counts with the aid of a flow cytometer.[13] Jon and Chris showed that our new method of turning brains into soup was not only at least as accurate as stereology, it was also faster, more reliable, and easier to apply.[14] And, as originally intended, it could provide numbers for complex, heterogeneous structures that could not be analyzed with stereology, like an entire brain.[15]

The new method, by the way, was not called "brain soup." It had been pointed out to us that our approach was akin to a liquid version of the optical fractionator, the stereological method used to fractionate tissue into slices, slices into blocks, and blocks into the contents of optical probes, whose cells were only then counted. In the same spirit, Roberto and I felt that our method should also be called a "fractionator." Since we were going beyond the cubes of tissue counted by the optical fractionator and breaking them into the ultimate units counted, nuclei, I suggested "ultimate fractionator," but Roberto wisely vetoed it. Because we were turning heterogeneous tissue into a homogeneous—or "isotropic"—suspension of nuclei, he proposed we call it the "isotropic fractionator." The name stuck for lack of any better alternative. It has been pointed out to me by none other than Karl Herrup himself that it's a terribly awkward name, and I agree. Whenever I can (which is not often because journal editors don't appreciate informality), I prefer to call our method of counting cells by what it is: "brain soup."

3 Got Brains?

Now that I had a method to count brain cells, I needed the brains—and much more than just human brains. I needed the brains of a broad range of other species to compare the human brain to: how else were we to find out whether ours had more neurons than even larger brains, like the elephant brain? I also wanted to understand the evolutionary origins of brain diversity: whether the relationship between brain size and number of neurons was the same for all mammals, whether different brain structures gained neurons at the same rate, whether there were any universal, fundamental laws that determined what brains were made of. Achieving that goal required the brains of many species as possible, small and very small, large and very large, from different mammalian groups, beyond the usual mouse, rat, and occasional monkey available in research labs.

But we had to start somewhere, of course, and it was convenient that our animal facility at the Institute of Biomedical Sciences in Rio had easy access not only to rats and mice, but also to hamsters and guinea pigs, four different species of rodents with an eightfold difference in body mass and a ninefold difference in brain mass. That was a good start: I felt that the few large-scale comparative studies of brain composition in the literature, those of Donald Tower and Herbert Haug in particular, had compared all sorts of mammals as if they were equals, and I didn't want to make the same mistake (Heinz Stephan and his team in Germany had been careful to separate primates from insectivores and bats in their studies, but, then again, they only had volumetric data, not even neuronal density estimates from which to ballpark numbers of neurons). Since we had easy access to four rodent species, I would stick to rodents for starters. But we needed more—and, specifically, we needed the brains of large

Figure 3.1
Capybara (*Hydrochoerus hydrochoerus*), the largest rodent species (left), and agouti (*Dasyprocta primnolopha*), the fourth-largest rodent species (right).

rodents if we were to find out the scaling relationships that applied to building the rodent brain.

We were in luck: the largest rodent was in South America—the capybara (*Hydrochoerus hydrochoeris*, for "water hog"; figure 3.1), a German shepherd–sized, square-faced, tailless brown furry mass that bears no obvious resemblance to rats and other rodents beyond the telltale huge incisor teeth. They are gregarious animals who wade in the margins of freshwater lakes and rivers in the Amazon basin and who hide underwater, nostrils alone above the surface, whenever large cats and snakes, their predators, appear. Luckily, too, capybaras were far from being an endangered species; they were considered food that was not only fit for human consumption but also tasty, at least in the northwestern parts of Brazil. Capybaras had been sighted in the lagoons of Rio de Janeiro, and they were becoming a plague in Campinas, not far away, in the state of São Paulo. That said, getting permission to trap and kill one was shaping up to be a bureaucratic nightmare. Besides, the Carioca capybaras had become dear to the general public, who celebrated their every sighting, and I had no intention of being the one to spoil the party. If I was going to be in the business of studying brain diversity for a while, I didn't want the bad press.

Just as I was starting to look for capybara farms that supplied restaurants in the Northwest, Roberto Lent, my collaborator, received news from Cristóvão Picanço-Diniz, a colleague at the Federal University in Belém do

Pará, in northern Brazil. The Brazilian Institute of Environment and Renewable Natural Sources (IBAMA) had seized two live capybaras raised illegally for food in a family's backyard and was about to dispose of them when it occurred to the officers to first ask Cristóvão if his lab, which studied sensory representations in the brain, had any use for them (no, I have no idea how the officers knew to ask him). Cristóvão was aware of our new foray into comparative brain studies and offered to obtain the capybaras and send us their brains. His students promptly donned butcher aprons, killed the two large animals, and airmailed us their heads in a large Styrofoam box, floating in paraformaldehyde. That was one macabre and stinky sight, yet one that we celebrated: we had capybara brains!

For good measure, Cristóvão also supplied us with two more brains of another large Amazonian rodent: the agouti (*Dasyprocta primnolopha*; figure 3.1). This is a housecat-sized, feisty animal who sits on its hind legs to eat food that it holds in its front paws, as rats do. As the fourth largest rodent, it was smaller than the North American beaver and the South American paca, but, with a body mass of 3–4 kilograms (7–9 pounds), it would certainly do. We now had a sample of six rodent species ranging in body size from 40 grams (1.4 ounces; the lab mouse, *Mus musculus*) to over 40 kilograms (90 pounds; the capybara), with brains ranging from 0.4 gram (1/70 ounce) to 75 grams (2.6 ounces, roughly comparable in mass to the brain of a macaque). All we had to do was turn them into soup.

Primate like Us

As a newcomer in the field, I was unable to gauge the real importance of the numbers of neurons and other cells that we were now able to determine in the different rodent species. I sensed we were onto something major, now that we were finally able to examine the cellular composition of different brains, but I wanted somebody else's opinion. Since we hadn't published anything yet on the isotropic fractionator method or the rodent brains, I needed to talk to somebody in person and explain what we were doing.

The opportunity arose in March of 2004, while attending the symposium that launched Miguel Nicolelis's International Institute of Neuroscience in Natal, northeastern Brazil. Jon Kaas, a leading expert in primate brain evolution and longtime friend of Miguel's, was to give a talk at the symposium.

I approached him right after his talk, a total stranger and nobody's disciple in the field, and asked him point blank: "What if I told you I had a very simple way to count neurons in whole brains, whole cortices, any structure of interest—how important would that be?" Jon opened his eyes wide and pulled his head back to peer at me through his tiny eyeglasses, as I now know is his custom when he's heard something surprising: "That's what we've been after all along, but nobody knows how to do that." That was all I wanted to hear. We were onto something useful at the very least.

Later that year, we presented our first results on rodent brains at the mecca of neuroscience, the annual meeting of the Society for Neuroscience (SfN) in the United States. Jon and his postdoctoral fellow Christine Collins came to our poster, and we started talking of a possible collaboration to investigate the numbers of neurons that composed the brains of the primate species they had easy access to in his lab at Vanderbilt University in Nashville. After our paper on the isotropic fractionator method came out in 2005, at the next SfN meeting, where Roberto and I were showing a second poster, on the changing numbers of neurons in the developing rat brain, Jon, Christine, and I met again. This time, the plans to take me to Vanderbilt to start collaborating on primate brains were more serious. Three months later, I was there.

It was January of 2006, and Christine Collins and a Ph.D. student, Peiyan Wong, had already arranged to have brains of several primate species hard-fixed—ready to be dissected and turned into soup upon my arrival. They had also organized all the necessary glassware and reagents and had reserved time on the microscopes, so we got down to work right away. Some three days later, I walked into Jon's office with my laptop open and showed him the first graphs comparing marmoset, galago, and owl monkey brains (which are all fairly small and easily turned into soup) to our by then complete (but still unpublished) rodent brain data set. The results were very promising: those three primate species seemed to pack a much larger number of neurons in their brains than rodents with a similar brain size. Their primate brains were clearly made differently from rodent brains. "So it actually works!" said Jon, eyes wide open, mouth agape, and head pulled back. I smiled, laptop still in hand. "You didn't expect it to work, did you?" He freely admitted to it. "But now you can have everything you want, anything you want. Just name it!"

And so the fun began. We set aside a number of brain hemispheres from different primate species, including several macaque species, that I would take home with me and analyze in my brand-new tiny lab in Rio. Jon introduced me to Ken Catania, his former student and biologist extraordinaire, also at Vanderbilt, who at the time had a colony of naked mole rats, ugly-looking tunnel-dwelling hairless rodents in his lab, and who also often trapped various eulipotyphlans (shrews and moles). So we started a collaboration on the brains of this group of animals, formerly known as "insectivores," which include some of the tiniest mammalian species around. Christine and Peiyan started dissecting visual and auditory structures in the cortex and subcortex of different brains so that we could investigate how the two functional pathways scaled in numbers of neurons. As primate brains increased in size, did vision take over hearing, as presumed in the highly visual primate species? (It turned out it did not, although we found that vision does take precedence: the nonhuman primate brain has around fifty times as many cortical neurons devoted to processing visual as to auditory information.[1]) I raided Jon's cold storage room, where I found a number of cerebellums and olfactory bulbs that had been sitting around as unwanted tissue. Jon's lab specialized in studying the cerebral cortex, but unlike most such labs, his had luckily not thrown these unwanted brain parts away. (The following year, when I came back with the full data set ready to be turned into a publishable paper, we would start collecting spinal cords from the animals that were killed in Jon's lab for other, unrelated studies—to give us a window on the relationship between body size and the number of neurons required to operate it.[2]) While rummaging in Jon's cold storage room that first visit, I found four large cerebellums, one from a gorilla and three from orangutans that had been sitting in a bucket of paraformaldehyde for over a decade. As a very stable molecule, the DNA should still have been inside the nuclei, so the DAPI stain that revealed the nuclei and allowed us to count them should still work, despite the long fixation time. Here was precious tissue we could put to good use. I ran to Jon's office. "May I please have them?" "Of course, you may." It was Christmas all over again.

Collaborating with Jon has been a life-changing and uplifting experience. He is a gentle, kind soul who has already accomplished an enormous amount scientifically, both in the fields of adult plasticity and of evolutionary neuroanatomy, and someone who now, in his own words, just "wants to have fun." He immediately sponsored the publication in the *Proceedings of the National Academy of Sciences* of our first study on rodent brains in 2006. He started spreading the word about our new data in comparative neuroanatomy circles, and he arranged for me to be the keynote speaker at the Karger symposium on Brain Evolution in 2010, where I finally got to know some of the key researchers in comparative and evolutionary neuroanatomy. I am certain, although I will never ask and he will never tell, that he also nominated me for the James S. McDonnell Foundation Scholar Award, which I won that year—an unprecedented (in Brazilian terms) amount of 600,000 U.S. dollars to be spent in pretty much any way I saw fit on our research toward understanding the neuroanatomical bases of the cognitive superiority of the human brain. My main achievements most certainly owe much to Jon's influence in the field—yet, to be fair, my findings wouldn't have been published or my achievements recognized if they weren't worthy of it. Jon has my unconditional gratitude, and to this day we continue to have fun exploring brain evolution, just as he has wanted. We have already published fourteen studies together, and more are yet to come.

African Mammals Large and Small

Primates were a major advance, and a necessary one if we were to determine whether humans have a standard primate brain. But I wanted a much broader range of species to tackle the mechanisms that generate evolutionary diversity. In that context, there was a name that had been mentioned to me a few times as a possible source of brains both very large and very small: Paul Manger, an Australian scientist in South Africa who focused on the comparative neurochemistry of brainstem structures—and who, I would later learn, loves to drive his Land Rover off road on the African savannas and do the collecting himself.

The time to contact Paul arrived when, in 2009, I came across a paper he had just published describing how to perfuse the brains of elephants in the wild and collect them still in good condition for microscopic neuroanatomical studies. Our report describing the average number of neurons in the human brain had just been published,[3] so we were ready to move on to the next question: did the much larger African elephant brain contain more or fewer neurons than the human brain? From reading his new paper, I knew that Paul had legally obtained the brains of three adult male African elephants, culled as part of a larger population management program undertaken in Zimbabwe by the Malilangwe Trust. Under the supervision of a wildlife veterinarian employed by the Trust, those three elephants had been darted with an overdose of an anesthetic safe for human consumption, then shot through the heart. The head of each elephant was separated from the body and perfused through the carotid arteries, first, with 100 liters of saline flowing from a raised platform to flush the brain of all blood and, then, with 100 liters of paraformaldehyde to preserve the brain immediately for transportation back to the lab and later studies. Meanwhile, the body of each elephant was butchered by another team, and its meat was distributed among the people living in the area, providing them with more than ten thousand meals. Those three animals, who would have been culled anyway along with several others that year, not only fed the locals, but also contributed to science, through Paul's own work and that of his ever-growing network of collaborators worldwide.

So I e-mailed Paul, wishing to join the club. I realized it was a long shot, writing to say what amounted to "Hello, stranger, I'm in the business of turning brains into soup to figure out what they're made of. Could I please have half of one of those brand-new elephant brains that you just got and destroy it?" Luckily, Paul didn't turn me down right away (as it turns out, he is actually very fond of unusual ideas, which must certainly have helped my case), and instead promptly wrote back offering to provide me with a few blocks of tissue that I could process. Would that be enough?

Thanks, but no, it would not be enough, I replied, and proceeded to explain that a few pieces would not allow us to go beyond what had already been done by Donald Tower in the 1950s, when he determined the density of neurons in a few samples of the enormous cerebral cortex of the Asian elephant.[4] Multiplying estimates of neuronal density in the cerebral cortex by the volume of the tissue would, in principle, be a simple mathematical way to estimate the total number of neurons in the entire tissue, were that entire tissue to be homogeneous. But, based on what we knew about the human brain, we couldn't assume that it would be. Nor was that calculation acceptable in the face of proper stereological methods that offered unbiased estimates, given that estimates of total numbers of neurons obtained that way for a large number of species would be hugely biased by the enormous variation in brain volume across species, so they would have no mathematical value for understanding how brains of different species compare. There was no way out. If we wanted to do the right thing and find an answer to the very important question of whether the human brain still contained more neurons than a brain three times as large such as the African elephant's (which I thought would be the simplest explanation for the cognitive superiority of humans), we needed to count neurons in a whole elephant brain. Or, at the very least, we needed a whole half brain—a hemisphere—assuming that any differences in numbers of cells across the two hemispheres would be negligible compared to large-scale variation across species.

Paul was happy with my sales pitch and immediately agreed to give me an elephant brain hemisphere! We started to work out the practical arrangements. The tissue was too precious to send through the mail and risk having it disappear or get stuck in the hands of Brazilian customs authorities. So Paul would apply for permission from the South African government to export the elephant brain hemisphere, and I would come visit his lab at the University of the Witwatersrand in Johannesburg and bring it home with me as accompanied luggage. But this raised another issue: could I actually do that?

I found the phone number for the National Agency for Sanitary Vigilance (ANVISA), which inspects Brazil's borders and is responsible for regulating the entry of food and health-related products into the country. I called the ANVISA office at Rio's International Airport, where I planned to land, and posed what had to be one of the most preposterous questions they had ever heard: "Hello, I'm a scientist going to visit a collaborator in South Africa, and I'm wondering, would it be okay to bring an elephant brain back with me in my suitcase?" But the lady on the other side of the line was nonplussed. "An elephant brain. Is it alive?" (What?) "No, ma'am, it's good and dead." "Then we don't care." As long as I had a list of the species whose tissue I was bringing into Brazil, and a declaration that the tissue was neither alive nor biologically dangerous nor of any commercial value, I was fine. What they really wanted to suppress was the trafficking of live animals. The brains I would be transporting were nowhere near that.

I visited Paul's lab in November of 2009 to collect the elephant brain hemisphere and ended up bringing an assortment of whole brains—of bats, African rodents, afrotherians (a wide range of mammal species living or originating in Africa), a giraffe, and an antelope—but no half brain from an elephant, at least not yet. We ran into a wall of bureaucracy trying to obtain the permits to export the elephant brain out of South Africa. All other brains were fine, though.

I would only get my hands on the elephant half brain in 2012, three years and many other brains later, when Paul and I organized our own foray into collecting the brains of a number of large artiodactyl species. Artiodactyls are even-toed ungulates like the pig, deer, and antelope, the largest of which are the giraffe (one brain of which Paul had given me on my first visit) and the hippopotamus (one brain of which is at this moment sitting in Paul's cold storage room waiting for me to go collect it). Artiodactyls are particularly interesting for a number of reasons, one being that they belong to the same mammalian order as the very large brained cetaceans (so the cellular composition of their brains could tell us what to expect from cetacean brains until I could actually get my hands on one): whales and dolphins share a common ancestor with the modern hippopotamus. The other reason is that artiodactyls have brains the size of medium and large nonhuman primate brains, so they could tell us whether primate brains had more neurons than artiodactyl brains of a similar size. That is, we would be able to compare a chimpanzee and a cow in terms of the number of brain neurons, not brain mass. If my hypothesis that the number of neurons was the main limitation to cognitive capability was right, then we should find far fewer neurons in artiodactyl brains than in primate brains of comparable mass.

Our chance to obtain those brains without having to hunt for the animals ourselves materialized thanks to a highly reputable business in South Africa, a maximum-security facility for wildlife seized (in the case of protected species injured by poachers) or caught legally in the wild, according to South African regulations, and quarantined until it could be safely returned to the wild or sold to foreign zoos, Arabian princes—or neuroscientists who just wanted the animals' brains. During a visit to plan another study on the continuously growing brain of the Nile crocodile, which had involved an excursion to a croc farm (again, perfectly legal),

Paul had taken me to the facility, where he had previously obtained and perfused giraffes. The entire area was double fenced and encircled in electrified wire: I felt like I was entering Jurassic Park and something wild was about to pounce on me. Several highly protected white rhinos were there, retrieved from poachers and recovering from their injuries until they could be safely returned to the wild. There we could also walk into a cheetah's enclosure, which we did, absentmindedly following the lead of the veterinarian who had raised her from a cub and swore that she was as tame as a pussycat. By the time my rational prefrontal cortex caught up, we were already too close to the cheetah, whose eyes were set on mine. Talk about a chill down my spine. But since running away would have been futile, I went ahead and petted the cheetah, who was indeed purring loudly, just like a kitty. That's right, I petted a cheetah. What an amazingly dumb thing to do.

I was given a (quite surreal) list of animals available and their prices and worked out with the fine people of the wildlife facility what we would like to do and how. Once back home, I went through the list looking up the respective body and brain masses of each species, made a sensible choice in terms of those species with a good variation in size, price, and number of hours required to collect their brains and asked both the James McDonnell Foundation and our university whether it was okay to buy their brains from the facility. It was. I would learn from the expert himself, Paul Manger, how to collect brains in the wild.

In June of 2012, we gathered a team of able-bodied, helpful students from Paul's department, several boxes of tools, many liters of saline and paraformaldehyde, and off we went. The arrangement was that the staff and veterinarians at the facility would euthanize the animals and allow us to do the dissections right there in the field; in return, they would keep the meat to feed their lions and other large cats, and their workers would cure and keep the furs for themselves. And we would also take care of cleaning up. Everybody was happy.

"Not the animals," my mother protested. "What a horrible thing to do, end their lives." Well, considering that everybody dies in the end, and that the alternative for these particular animals would have been to die horrible, gory, painful deaths in the wild from being gutted alive by lions and leopards, I think that dying a painless death from an overdose of anesthetic was actually not so bad. But, then again, I'm a happy carnivore (so is my mom, by the way, though she, like so many people, refuses to eat "cute animals"). I am also an animal and, by definition, animals have to feed on other life forms. Lions also have to eat. We just happened to be the ones to kill *their* dinner for them that day.

Paul set up with some students to remove the brains; I set up with others to remove the spinal cords—which, I must say, was a lot more work. We started by removing extra-large strips of filet mignon from the back of each animal with some really big knives. I will just say that I found it strangely satisfying that I could learn so quickly how to do precision work with a chain saw,* so that by the time we were working on the last of twelve animals, we had our system down pat and were even beating Paul's team (to the great satisfaction of this newbie).

Back in the lab, I prepared two suitcases full of sealed containers double wrapped in white plastic with one or more brains each. I sent home a picture of me and my bounty. My husband's reply was swift: "It looks like you're bringing home bales of cocaine. I'll go get the bail money ready for when they arrest you at the airport."

But the lady from ANVISA was true to her word. Sure enough, I was stopped at customs after they X-rayed my suitcases—but only because the agents thought I was bringing highly prohibited fresh cheese, like the Portuguese couple they'd scanned just before me. "No, it's only brains for a research collaboration with a university in South Africa." It was so unusual a situation that they didn't know who exactly should inspect my luggage and papers. So I sat there and waited patiently. The lady officer who came to do the inspection read the declaration of nonhazardous biological materials with no commercial value, browsed through the pile of permits (I had one for each species, written in several different languages), glanced through the long list of species, kept the copy of the list I knew she would demand from me and then … let me and my brains pass on through. No bail was needed.

As the papers got published and prospective collaborators became interested in our work, it became easier and easier to find the brains we wanted

*I sometimes toy with the idea of the bizarre entries I could make in a new "Other skills" section of my CV: "Knows how to turn brains into soup; has a weird knack for spotting wrong entries in her students' data tables; can prepare filet mignon from kudu, eland, and crocodiles; can do precision work with a chain saw."

to analyze. We now have the brains of marsupials, carnivores, birds, fish, octopodes, and, finally, cetaceans in the works. I would already be working on insects if I knew how to dissect their brains (yes, they do have brains): not only are insects the most diverse group of animals with the greatest number of species, they are also readily available in any backyard. And since people usually have no qualms about squashing cockroaches with their shoes or swatting mosquitoes with their hands, I could expect little resistance or protests of disgust from the public. My main interest is brain diversity and what it teaches us about how life evolved. So if it has a brain, I'm interested. It says so on a black sticker on the door to my small lab: *Got brains?*

4 Not All Brains Are Made the Same

In a nutshell, what I wanted to find out was what the brains of different species were made of, which to me was one of the most fundamental questions of neuroscience—but one about which we still knew very little at the time. We knew, of course, what kinds of cells brains were made of—neurons, glial cells, and the endothelial cells that formed the walls of the capillaries that brought oxygen and nutrients to the brain through the blood—but in what quantities, and in what proportions? What were the rules that determined how brains were built, if there were any at all? Did increasing brain size across adult brains (whether measured as mass or volume, which here are interchangeable*) simply mean adding proportionate numbers of these cells, or were different brains built differently, with different proportions of each cell type? Essentially, was there a single relationship between brain size and its number of neurons and other cells, one that applied universally across all species? That is, were all brains made the same way?

If they were, as it was supposed when I got into the business of counting cells, then brains of similar mass should have similar numbers of neurons even if they belonged to widely different species; and the larger the brain, the more neurons it should have, even across entirely unrelated species. That meant there was a simple test to the hypothesis: there should be a single relationship between the mass of a brain and the number of neurons in it across all mammalian species—and two brains of similar size should always be made of similar numbers of neurons.

*Brain volume and brain mass are essentially interchangeable, with a proportionality factor of 1.036. More specifically, brain volume equals brain mass divided by a factor of 1.036 g/cm^3, which is the specific density of the brain (just slightly larger than pure water, which by definition has a density of 1 g/cm^3).

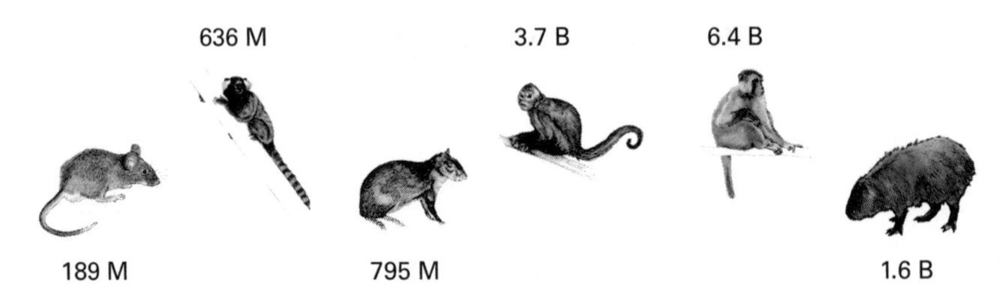

Figure 4.1
Rat, marmoset, agouti, owl monkey, rhesus monkey and capybara lined up by brain mass. The two monkey species have more neurons than the capybara, despite their smaller brains. Numbers of neurons are indicated (M, million; B, billion).

In 2007, we had enough data from six species of rodents and six species of primates (figure 4.1) to put that hypothesis to the test.* We knew for the first time that, for example, the brain of the rat had, on average, 189 million neurons; the agouti brain, 795 million neurons; and the much larger capybara brain, 1.6 billion neurons.[1] Across primates, working with Jon Kaas, we found, in contrast, 636 million neurons in the marmoset brain, 3.7 billion neurons in the capuchin monkey brain, and 6.4 billion neurons in the rhesus monkey brain.[2] These simple numbers already show that midsized primates had many more neurons than even the largest rodent species—but sizes, and appearances, can be deceiving. This is where mathematical analyses and statistics become fundamental: so that we need not rely on appearances.

A simple, linear plot of brain mass as a function of the number of neurons in the brain of each of these first rodent and primate species (figure 4.2) shows that, overall, the more the neurons a brain has, the larger that brain is: larger brains do tend to have more neurons. But at this point, the question remained: did all brains become larger by the same amount as they gained neurons?

In the linear plot in figure 4.2, it is difficult to distinguish among the smallest rodent and primate species, which appear bunched up at the bottom left corner. It appears that a single relationship applies to all these species—except to the capybara, which stands out to the left of the larger

*A list of species, brain mass and number of neurons in our dataset is found in the appendix. The complete dataset is provided in Herculano-Houzel, Catania, Manger, and Kaas, 2015, and is also available at www.suzanaherculanohouzel.com/lab/.

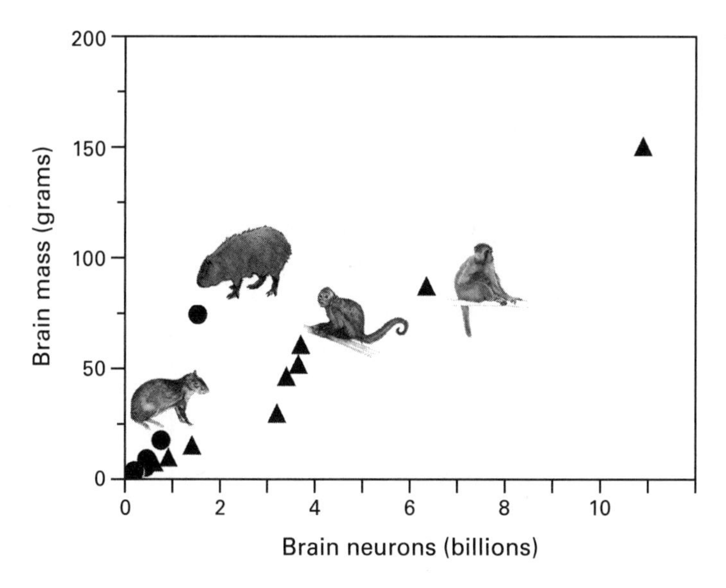

Figure 4.2
Linear plot shows what could still be a single relationship between brain mass and the number of neurons in the brain across rodents (circles) and primates (triangles), making the capybara appear to be an outlier.

primates as if it were an outlier, with an exceptionally made brain, too large for its number of neurons. Any appearance of ambiguity can be dispelled, however, by making the plot double-log (log-log; figure 4.3), that is, by using a log scale on both horizontal and vertical axes. As shown in chapter 1, this is a standard maneuver in quantitative biology that makes the smallest values appear more separated, as the log-log scales turn into a straight line a power law that appears in a linear scale as a rapidly upward-sloping curve. The result is equivalent to plotting log values in a linear scale, with the advantage that the graph shows the nontransformed values directly.*

*The standard way of using power laws in studies of allometry is to use body mass as the independent variable, or *X*. Similarly, the tradition in studies of how brain morphology varies with brain mass is to use the brain mass as *X* on the graphs. But because brain mass can only be a result of the number of neurons and their average size, defined by a set of mechanisms in development, I decided that the logical thing to do was to always plot brain mass as the dependent variable (the consequence of number of neurons), that is, as *Y*.

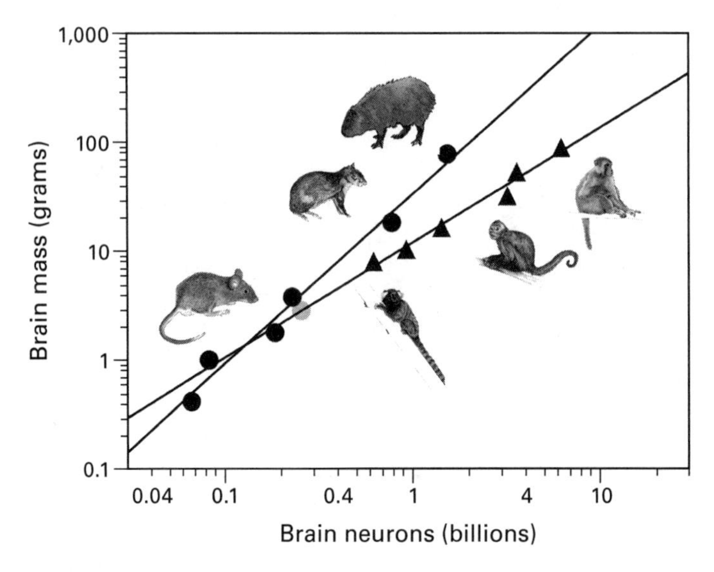

Figure 4.3
Double-log plot shows that different relationships exist between brain mass and the number of neurons in the brain for rodents (circles) and primates (triangles). The lines indicate the power laws that best describe the variation in brain mass as functions of numbers of neurons in the brain of rodents and primates, separately, with exponents of +1.6 in rodents and +1.0 in primates. Now the capybara no longer appears to be an outlier: it has the brain mass predicted for a rodent brain with its number of neurons.

The double-log graph in figure 4.3 suggests that rodents and primates are actually clustered along two separate lines—and that the capybara is, in fact, no outlier, at least not when it is compared to other rodents. As pointed out in chapter 1, data points forming a line in a double-log graph are indicative of a power law relationship of the kind where Y (vertical axis), instead of varying proportionately with X (horizontal axis) multiplied by a given constant number, varies with X *raised* to a constant power (a). Power laws are thus written as $Y \sim X^a$, in contrast to linear functions, where $Y \sim cX$. In a way, a linear function is a special case of a power law with the allometric exponent $a = +1$, that is, where Y increases only linearly with X. But as soon as the exponent a becomes significantly larger than 1, Y begins to increase faster than X. For example, when Y is brain mass, X is the number of brain neurons, and the power law relating them has the allometric exponent +1, then the relationship is linear, and a 10 times larger number of neurons results in a brain exactly 10 times larger. But when the allometric exponent of the power law is +2, then the same 10 times larger number of neurons results in a brain 100 times larger.

In our first analysis,[3] which did not include the human brain, we found that brain mass in primates scales (increases) with the number of neurons in the brain raised to the power of +1.0—that is, the primate brain scales in an essentially linear manner as it gains neurons across species. Indeed, at 87 grams, the brain of the rhesus monkey is just over 10 times as large as the marmoset brain (which weighs only 7.8 grams) and has close to 10 times as many neurons. In contrast, at 17.6 grams, the brain of the agouti is about 10 times as large as the rat brain, but has only about 4 times as many neurons. This, we found, is because brain mass scales much faster as rodent brains gain neurons,[4] as a power law with allometric exponent +1.6. With this exponent, a ten times larger number of neurons results in a rodent brain that is not just ten times, but rather forty times larger in mass. In comparison to primates, therefore, the way that rodent brains gain neurons is inflationary: it leads to a rapidly ballooning brain mass. As a consequence, primate brains not only have more neurons than rodent brains of a similar mass, but the larger the mass of the brain, the larger the difference in the number of neurons found in a primate versus a rodent brain, shown as the progressive separation between the two plotted lines in figure 4.3. There was our answer, then: not all brains were made the same, for even the evolutionarily close cousins rodents and primates had very different numbers of neurons composing brains of comparable size.

Because the brain consists of structures of very different layout, composition, and function, throwing all these together into one single "whole brain" lump might result in missing similarities or differences in one part of the brain across different animal groups. On the other hand, dividing the brain into too many structures would enormously increase the time necessary to gather data. For the initial studies aimed simply at establishing whether all brains scaled the same way across mammalian groups, we therefore decided to separate the brain into only the most obvious structures: cerebral cortex, cerebellum, and the anatomically inappropriate yet accurate "rest of brain." This did not include the olfactory bulbs, which, lodged as they are in a small bony case in front of the brain, are often left behind when researchers collect the brain. Since many brains we would receive for analysis would be missing the olfactory bulbs, to ensure that all brains analyzed would be comparable, we decided simply not to include the olfactory bulb in the total brain mass.

Indeed, we found that neurons were distributed in very different numbers across brain structures. Still considering for the moment only rodents and nonhuman primates, although already in a larger sample of twenty-two species,[5] we found that, in all of them, the vast majority of brain neurons, about 80 percent, were located in the relatively small cerebellum, in the back of the brain.[6] It was therefore possible that the different relationships that applied between brain mass and number of neurons for rodents and primates were mostly due to differences in the cerebellum, and not in the cerebral cortex, which might actually scale similarly across rodents and primates.

But that was not the case: changing numbers of neurons corresponded to very different masses of the cerebral cortex alone between the two groups. We started referring to these different relationships between the mass of a brain structure and its number of neurons across species as the "neuronal scaling rules" that applied to each part of the brain within each mammalian group. Across rodents, we found that the cerebral cortex gained mass according to a power law of its number of neurons with a large allometric exponent of +1.7; across primates, the exponent was +1.0, and therefore the scaling was linear. As seen in figure 4.4, this meant that, just as the whole brain did, the cerebral cortex itself scaled much differently between rodents and primates. As both gained similar numbers of neurons, the rodent cortex became much larger than the primate cortex—and, as a result, a rodent cortex had fewer neurons than a primate cortex of similar mass.

For example, the cerebral cortex has similar mass in the capybara (48.2 grams) and in the bonnet monkey (48.3 grams), but these similar values in mass conceal a highly significant difference in number of neurons: whereas the capybara cortex has only 306 million neurons, the bonnet monkey cortex has 1.7 billion neurons, that is, nearly six times as many neurons packed more tightly in a similar volume. This difference reflects what I have called the "primate advantage":[7] because of the way a primate brain is put together,

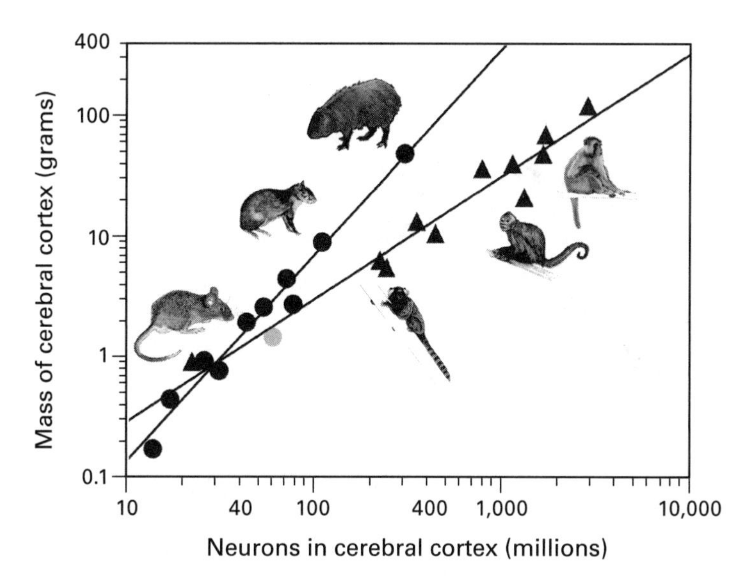

Figure 4.4
The cerebral cortex scales differently in mass between rodents (circles) and primates (triangles) as it gains neurons. The lines indicate the power laws that best describe the variation in cortical mass as functions of numbers of neurons in the cerebral cortex of rodents (exponent, +1.7) and primates (exponent, +1.0), separately. The larger the cerebral cortex, the larger the discrepancy in number of neurons found in primate and rodent species, with more and more neurons in primate than in rodent cortices.

a much larger number of neurons fits into a cerebral cortex of similar size to that of a rodent—which is a major asset when volume is at a premium, as we'll see later in this chapter.

The primate advantage is also seen in the cerebellum, which also turned out to have different neuronal scaling rules across rodents and primates. The cerebellum gains mass across rodent species according to a power law for its number of neurons with an allometric exponent of +1.3 (figure 4.5); across primates, however, the exponent is +1.0, and the scaling therefore linear. This means that the cerebellum scales differently between rodents and primates, with rodent cerebellums becoming larger faster than primate cerebellums as both gain similar numbers of neurons. As a result, more neurons fit in the cerebellum of a primate than in a rodent cerebellum of similar size. For example, the cerebellum of the capybara (6.6 grams) has 1.2 billion neurons, whereas the slightly smaller cerebellum of the bonnet

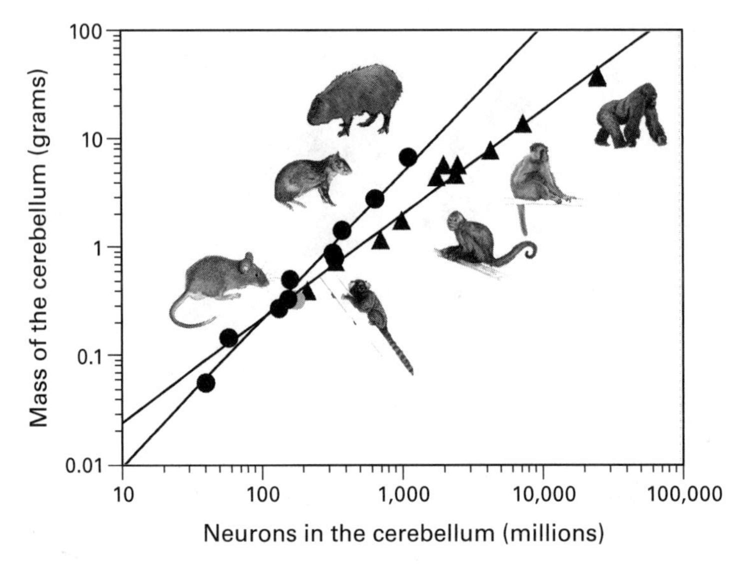

Figure 4.5
The cerebellum also scales differently in mass between rodents (circles) and primates (triangles) as it gains neurons. The lines indicate the power laws that best describe the variation in cerebellar mass as functions of numbers of neurons in the brain of rodents (exponent, +1.3) and primates (exponent, +1.0), separately. The larger the cerebellum, the larger the discrepancy in number of neurons found in primate and rodent species, with more and more neurons in primate than in rodent cerebellums.

monkey (5.7 grams) has 2.0 billion neurons, that is, almost twice as many neurons.

In contrast, the other brain structures, which we called the "rest of brain" and processed together as one lump of tissue, did not at first appear to scale so differently across rodents and primates. As seen in figure 4.6, this ensemble of brainstem, diencephalon, and striatum (all subcortical structures lumped together) gains mass across rodent species as its number of neurons increases according to a power law with an allometric exponent of +1.6, which is nominally different from the exponent +1.1 for the power law relating rest of brain mass to number of neurons across primates, but shows great overlap across the two groups. In contrast to the tidy alignment of data points for the cortex or cerebellum within rodents or primates, there is a much wider spread of the species around the fitted functions for the rest of brain, possibly because of the hodgepodge of different brain structures thrown into the same pot. Several primate species here fall on the function line plotted for rodents, and vice versa, as seen in the mixing of circles and triangles in figure 4.6. For instance, the rest of brain of the agouti weighs 6.0 grams and has 49 million neurons, not very different from the rest of brain of the bonnet monkey, which weighs 7.4 grams and has 61 million neurons. Scaling rules for the rest of brain, therefore, did not at first appear to be systematically different between rodents and primates. As we analyzed more species, however, it became clear that the primate advantage also applies to how its "rest of brain" is put together, as we'll see later. But for now, let's stick to what the primate advantage means for how its cerebral cortex and cerebellum scale as they gain neurons.

The different neuronal scaling rules for cerebral cortex and cerebellum across rodents and primates have very interesting implications in terms of the evolution of brain diversity. For the first time, we could suggest that

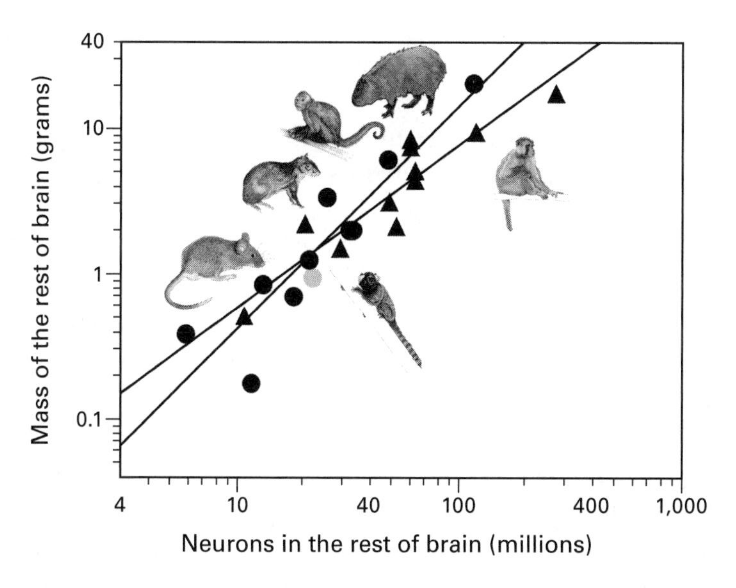

Figure 4.6
The rest of brain, in contrast to the cerebral cortex and cerebellum, initially appeared to scale fairly similarly in mass between rodents (circles) and primates (triangles) as it gains neurons. The lines indicate the power laws that best describe the variation in rest of brain mass as functions of numbers of neurons in the brain of rodents (exponent, +1.6) and primates (exponent, +1.1), separately. Although the exponents are nominally different, there is significant overlap between rodents and primates.

those events which led to the divergence of primates from the common ancestor that modern primates shared with modern rodents included not only changes in the positioning of the eyes (frontal in primates; lateral in most rodents) and in the termination of digits (finger and toenails in primates and claws in rodents), but also changes in how the cerebral cortex and cerebellum were built. The larger allometric exponents for these structures in rodents suggested that cortical and cerebellar neurons became, on average, larger in the rodent than in the primate lineage as these neurons also became more numerous (and larger in this case means not only a larger cell body but also longer dendrites and axons, as we'll see later). In contrast, the linear scaling of cortical and cerebellar mass as these structures gain neurons in the primate lineage indicates that the average size of their neurons must remain fairly stable across species. Whatever changes led to the divergence between rodent and primate lineages in evolution must therefore have included changes in genes and other mechanisms that control cell size.

Finding that the rodent and primate cerebral cortex and cerebellum had different neuronal scaling rules was extremely important. It showed that the widespread assumption in the literature that comparisons of brain size across widely different species could inform about their numbers of neurons and thus also about their cognitive capabilities was just plain wrong. Brains of similar sizes could be made of very different numbers of neurons—at least across rodents and primates.

The different neuronal scaling rules did not, however, tell us anything about whether the brain of the last common ancestor between rodents and primates (1) had been built like a modern rodent brain, and primates diverged away from it; (2) had been built like a modern primate brain, and rodents diverged away from it; or (3) had been built according to completely different scaling rules from those governing either modern rodent or primate brains—and *both* rodents and primates diverged away from it, each group with its own particular ways of putting together a cerebral cortex and cerebellum. On the other hand, the similarity in the neuronal scaling rules for the rest of brain that applied to rodents and primates at that point raised the possibility that the rest of brain of the last common ancestor to the two lineages was built in the same way as that in modern animals of these groups, a far more parsimonious scenario than that the two lineages had independently converged to arrive at the same neuronal scaling rules for the rest of brain—although even this scenario was, of course, still possible.

But there was a simple way out of this quandary. If we could determine the neuronal scaling rules that applied to the cerebral cortex, cerebellum, and rest of brain of other groups of mammals ranging from afrotherians to artiodactyls—mammalian species far more distant in their evolutionary histories than the closely related rodents and primates—we should also be able to determine which of all these possibilities was most likely.

Beyond Rodents and Primates

Living primates and rodents are close evolutionary cousins, related just as each of us and our first cousins are related—and not because primates descended from rodents or we descended from our cousins (which we obviously didn't, seeing as we have similar ages), but because both they and we descended from a shared common ancestor. In the case of our first cousins and us, this common ancestor is a grandparent, who may no longer be alive, but we can determine from oral history and written records when he or she was alive. In the case of rodents and primates, the last ancestor in common, and shared exclusively by the two groups and no others, lived some 95 million years ago.[8]

Such genealogical relationships with ancestors that lived and died so long ago can be established thanks to several modern and not-so-modern techniques and methodologies. Whether efforts to establish these relationships involve radiometric dating, gene and protein sequencing, or even the old-fashioned comparison of morphological characteristics, they are all subject to the mathematical analyses of "cladistics," tasked with determining which groups of species (clades) descend from a common ancestor and which do not. Of course, cladistic analyses don't always agree; different results can be expected depending on the precise species, genes, and proteins compared, because of different mutation rates over time among other reasons. The generally accepted genealogical tree of living species keeps evolving, changing as new evidence comes in that challenges former interpretations of the data and suggests new theories of evolution.

Still, it is a fact that the species that roam the Earth today were not always there, just as it is equally a fact that there once were species on the planet that are no longer around. Skeletons recognizable as belonging to modern humans are no more than 200 thousand years old; trilobites, on the other hand, thrived between 540 and 250 million years ago, but are not found in the fossil record either before or after that period. The same goes for dinosaurs: found embedded in rocks 230 million years old, but no older, they vanished 65 million years ago, at about the time a large meteor or comet struck the Yucatán peninsula in Mexico.[9] Ever since, no live dinosaurs have been sighted—although modern lizards and crocodiles bear an uncanny resemblance to them. Life, as preserved in the fossil record, has undoubtedly evolved over time—for "evolution" means simply change. Evolution is thus a fact, not a theory: life *has* changed over time. On the other hand, accounts of *how* exactly those changes have happened, through what mechanisms, leaving what paths behind to be traced as their histories are indeed theories.

Current consensus[10] based on modern molecular techniques and parsimonious mathematical analyses has it that, of all the modern eutherians—(placental mammals, that is, not monotremes and marsupials) —afrotherians diverged the earliest from the ancestor common to all, and cetartiodactylans (artiodactyls and cetaceans) diverged last or most recently, as illustrated in figure 4.7. Rodents and primates lay in between, lumped together in the supergroup Euarchontoglires—a lumping we had already showed, in 2007, not to be valid in terms of how their brains were built. If we could determine the neuronal scaling rules that applied to afrotherians, on the one hand, and to artiodactyls, on the other, and compare them with the scaling rules for rodents and primates (and perhaps even other clades as well), we should be able to tease out the most parsimonious explanation for how differently built brains came to be as they are today.

By 2014, around ten years after creating the method of turning brains into soup to find out how many cells they were made of (well, *had* been made of), and thanks to a small army of students and collaborators, we had published data for forty-one species spread across six different mammalian clades. Together with Ken Catania, we were able to determine the scaling rules that apply to eulipotyphlans, some of the smallest mammals alive; in collaboration with Paul Manger, we now knew the scaling rules for

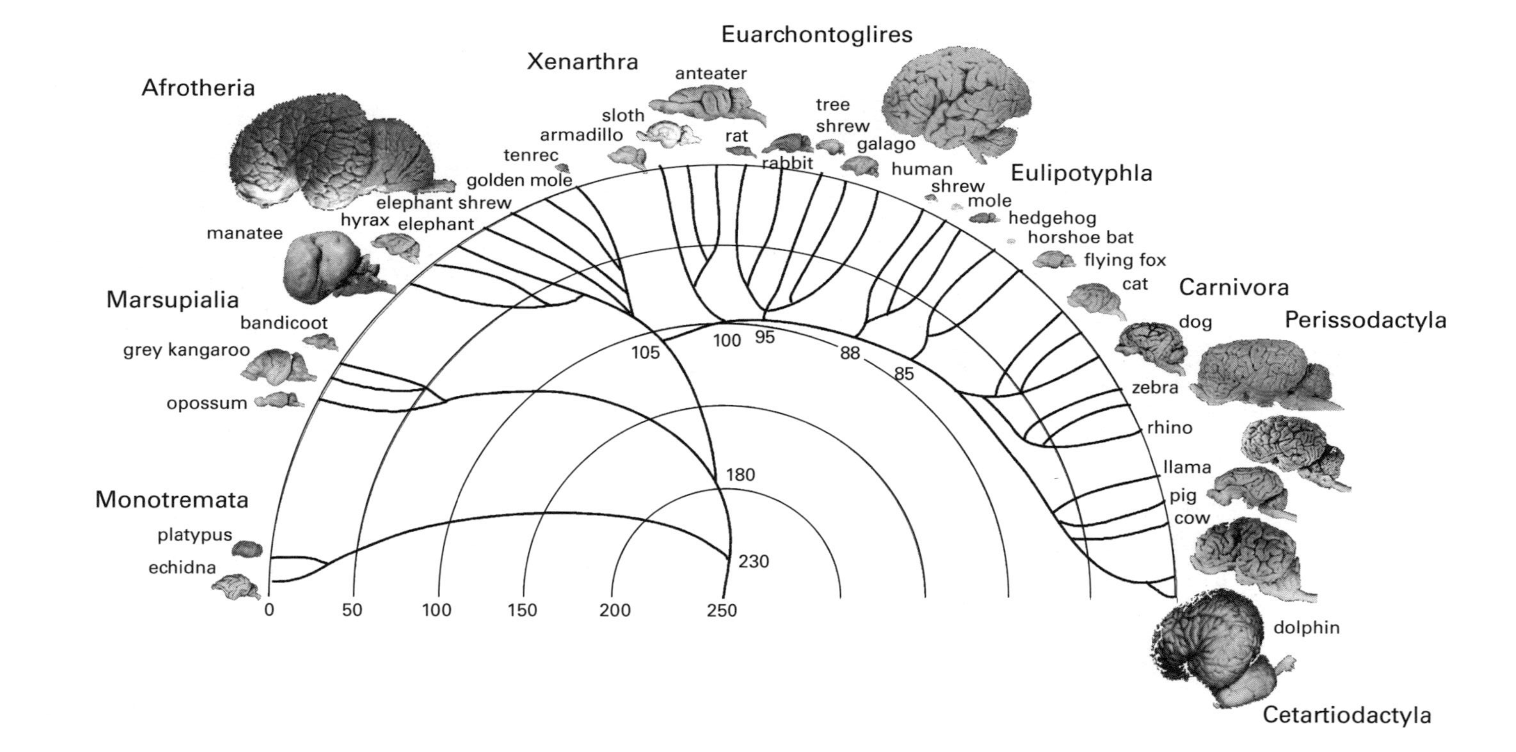

Figure 4.7

Consensus tree of the genealogical or evolutionary relationships among living mammalian species (dates are in millions of years ago). Rodents and primates are different branches of the same group, Euarchontoglires—and yet, different neuronal scaling rules apply to their cerebral cortex and cerebellum. Figure taken, with permission, from Herculano-Houzel, 2012.

afrotherians and artiodactyls, on opposing ends of the eutherian evolutionary tree. We could now examine all these species and look for differences and similarities in how their brains scaled in mass as they gained neurons, and thus get our first glimpse into the makings of brain diversity in mammalian evolution.

We found that the picture for the cerebral cortex was pretty clear. The neuronal scaling rules we had seen previously to apply to rodents also extended to the very old group of afrotherians (shrews and moles, plus the African elephant); to the very small but not so old eulipotyphlans (more moles and shrews), by then recognized as a separate group from afrotherians (thus retiring the previously used name "insectivores"[11]); and to the much more recent group of artiodactyls (pig, kudu, giraffe*). As shown in figure 4.8, all nonprimate cerebral cortices share the same relationship between cortical mass and number of neurons, while all primate cortices share another, different relationship among them.

We had an answer to our conundrum, then, illustrated in figure 4.9. Given the evolutionary relationships across the mammalian species and groups we were examining, the most parsimonious explanation for the nonprimate and primate scaling rules that we found was as follows. The nonprimate scaling rule, shared across modern nonprimate species belonging to mammalian clades young and old in their evolutionary divergence, must have been the ancestral way of putting together a eutherian cerebral cortex and remained fairly unchanged since then (otherwise, each of the nonprimate groups we examined would have had its own scaling relationship). Primates, in turn, had diverged away from that ancestral scaling rule—and found a way—the primate scaling rule—to pack more neurons together in a similar volume.[12] (The burning question, of course, was how humans compared to other primates—but that will have to wait until the next chapter.)

*Unfortunately, the giraffe brain that we could analyze belonged to a juvenile, not to a fully grown adult. But because we know that in rats and mice the total number of neurons in the cortex is attained well before the cortical mass reaches adult values, we can reasonably expect that the number of neurons that we found in the juvenile giraffe cerebral cortex does indeed represent the full adult complement, even if that cortex is not yet fully grown in size. For this reason, the giraffe is shown in the plot, but is excluded from the calculation of the power law in figure 4.7

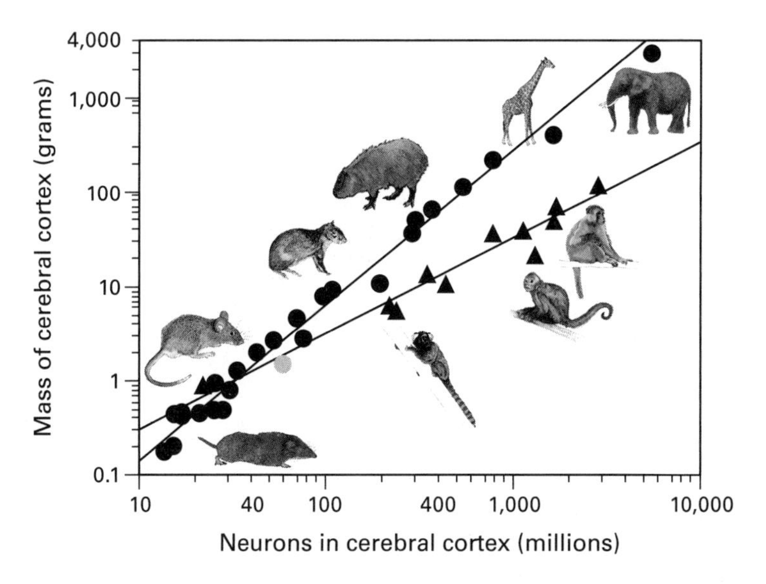

Figure 4.8
The cerebral cortex scales similarly in mass across rodents, afrotherians, eulipotyphlans, and artiodactyls (circles) as it gains neurons, but differently across primates (triangles). The lines indicate the power laws that best describe the variation in cortical mass as functions of numbers of neurons in the cortex of primates (exponent, +1.0) and of all other clades (exponent, +1.6). The larger the cerebral cortex, the larger the discrepancy in number of neurons found in primate and nonprimate cortices of similar mass, with more and more neurons in the cortices of primates than in those of nonprimates.

We can speak of primates' "diverging" away from the ancestral scaling rule as the cerebral cortex became larger because the evolutionary history of mammalian species is one of a very strong trend toward increasing brain size,[13] beginning with small animals with very small brains. The closest known relative to all modern mammals is the mammaliaform *Hadrocordium wui*, with an estimated body mass of only 2 grams, similar to the smallest living eulipotyphlan and bat,[14] that lived some 195 million years ago.[15] *Hadrocordium* had an estimated brain mass of 0.04 gram, with a cerebral cortex that we predicted to weigh 0.02 gram and to have had only 3.6 million neurons,[16] which is a far smaller cortex with far fewer neurons than we found in the modern mammalian species we sampled. But we can envisage that those early species had a cerebral cortex that was

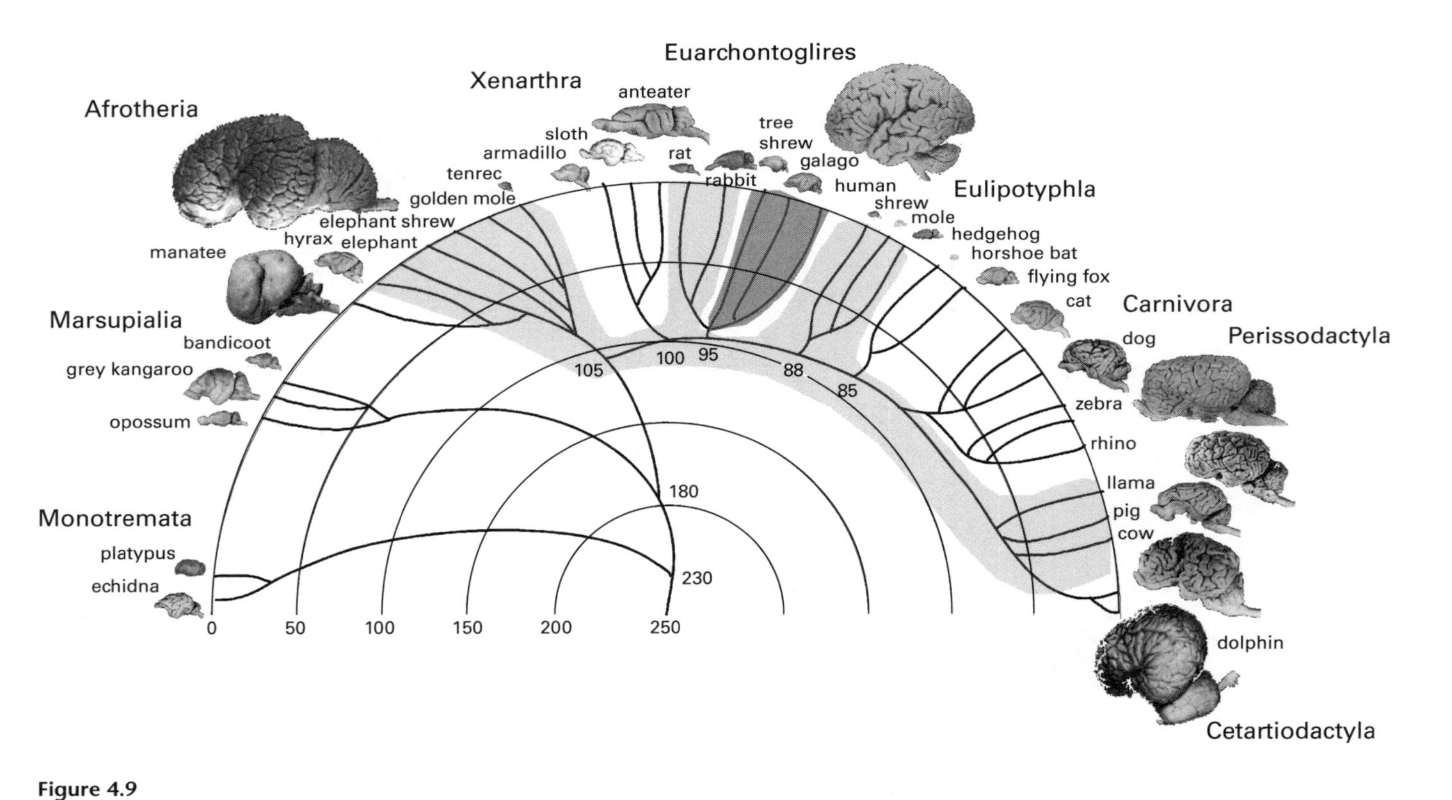

Figure 4.9
Proposed scheme for the evolution of the scaling of the cerebral cortical mass with increasing numbers of neurons: the neuronal scaling rules that apply to modern afrotherians, rodents, eulipotyphlans, and artiodactyls are presumed to already have applied to their common ancestor, to have been maintained in the evolution of these lineages, but to have changed in the divergence of the animals that later were found to have given rise to primates.

already built according to neuronal scaling rules that have remained the same in the lineages that gave rise to modern afrotherians, rodents, eulipotyphlans, and artiodactyls alike—and the group that eventually led to modern primates branched off as the way neurons were added to the cortex changed, giving primates the advantage of packing more neurons in a similar volume.

What about the cerebellum? Here we found that afrotherians (other than the elephant, as we'll see in chapter 6) and artiodactyls also shared the same neuronal scaling rule we had earlier found to apply to rodents, whereas eulipotyphlans differed from both primates and the other nonprimate groups, as shown in figure 4.10. It is interesting to note that, like the primate cerebellum, the eulipotyphlan cerebellum is packed with more

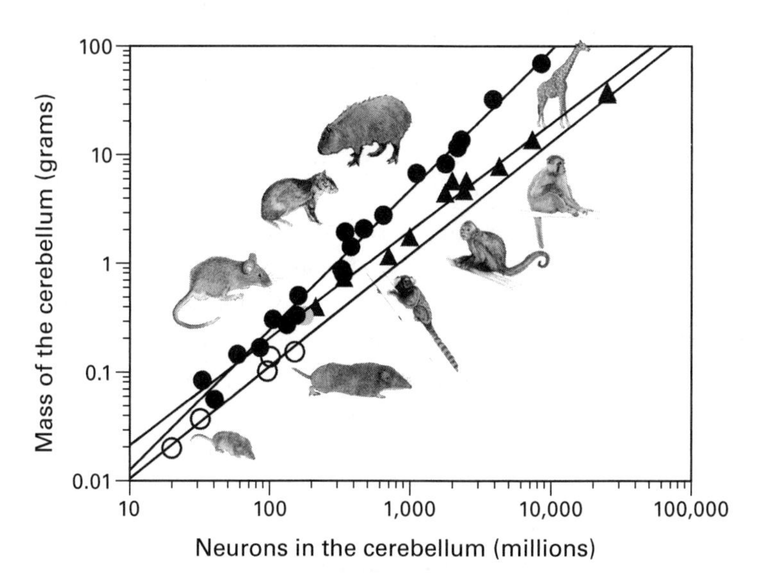

Figure 4.10
The cerebellum scales similarly in mass across rodents, afrotherians, and artiodactyls (filled circles) as it gains neurons, but differently across primates (triangles) and eulipotyphlans (open circles) as these gain neurons in the cerebellum. The lines indicate the power laws that best describe the variation in cortical mass as functions of numbers of neurons in the cortex of primates (exponent, +1.0), eulipotyphlans (exponent, also +1.0, but with a vertical offset in the graph), and all other clades (exponent, +1.3). The numbers of cerebellar neurons found in eulipotyphlans, though comparable to those found in small rodents and afrotherians, are packed into smaller volumes.

neurons than either a rodent or an afrotherian cerebellum of comparable mass. For instance, the cerebellum of the eastern mole (a eulipotyphlan), at 0.153 gram, has 158 million neurons, whereas the slightly smaller cerebellum of the hamster (a rodent), at 0.145 gram, has only 61 million and the cerebellum of the elephant shrew (an afrotherian), although slightly larger at 0.168 gram, has only 89 million neurons.

Thus it appears that the early eutherian species had a cerebellum that was built according to neuronal scaling rules that have remained the same in the lineages that gave rise to modern afrotherians, rodents, and artiodactyls alike—and the groups that eventually led to modern primates, and separately to eulipotyphlans, branched off as something changed in the way that neurons were added to the cerebellum (figure 4.11), allowing it to increase in mass more slowly as it gained neurons in both eulipotyphlans and primates, which then had the advantage over other mammals in how their cerebellums managed to pack more neurons without becoming disproportionately larger.

The rest of brain showed the same pattern as the cerebral cortex, with neuronal scaling rules shared across afrotherians, rodents, eulipotyphlans, and artiodactyls, indicating that these were the rules according to which the rest of brain of the first eutherians was also built. Primates, once more, diverged away from the common, ancestral way of putting together the rest of the brain with similar numbers of neurons packed into a smaller volume (figure 4.12). For example, having between 106 and 122 million neurons, the rest of brain weighs 20.0 grams in the capybara (a rodent), 64.7 grams in the kudu (an artiodactyl), but only 9.2 grams in the rhesus monkey (a primate).

What the Different Neuronal Scaling Rules Mean

That there was not a single, universal relationship between the mass of brain structures and their number of neurons was both a novel and a fundamental finding. We now knew that like must be compared to like—that humans, as primates, must only be compared to other primates, not to rodents or artiodactyls. The mass of the cerebral cortex, for instance, was a useful proxy for the number of cortical neurons when comparing rodents to other rodents or even to artiodactyls—but primates could no longer be compared to rodents and artiodactyls as if they were equals, only to other

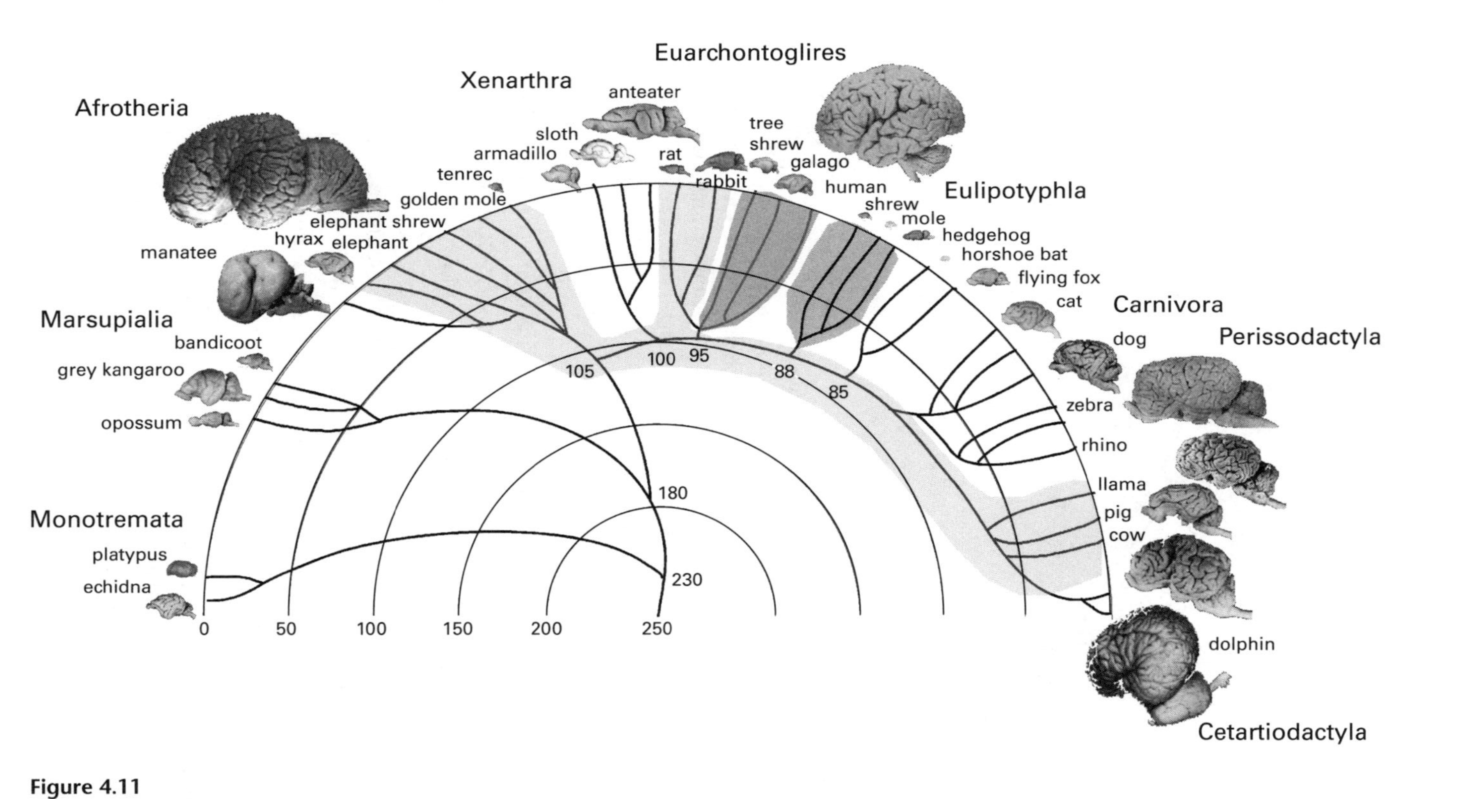

Figure 4.11
Proposed scheme for the evolution of the scaling of cerebellar mass with increasing numbers of neurons: the neuronal scaling rules that apply to modern afrotherians, rodents and artiodactyls are presumed to already have applied to their common ancestor, and to have been maintained in the evolution of these lineages, but changed twice, and separately, in the divergence of the animals that later were found to have given rise to primates and to eulipotyphlans.

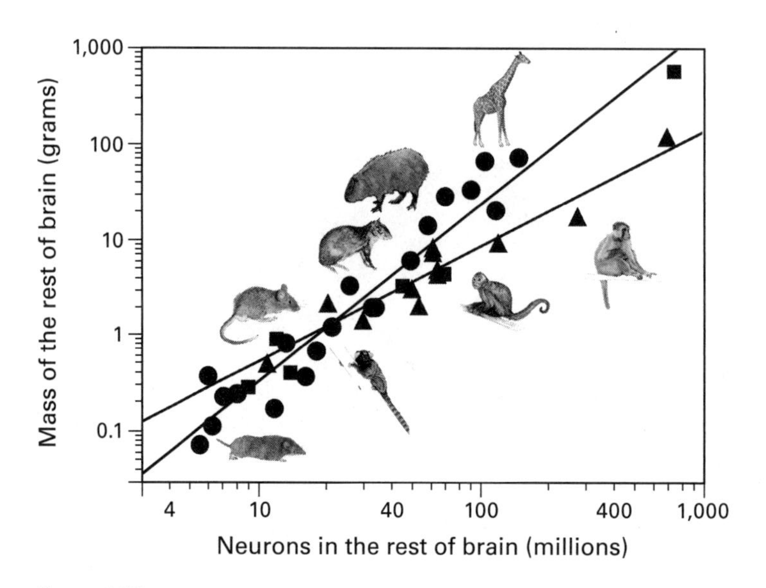

Figure 4.12
The rest of brain scales similarly in mass across afrotherians (squares), rodents, eulipotyphlans and artiodactyls (circles) as it gains neurons, but differently across primates (triangles). The lines indicate the power laws that best describe the variation in the mass of the rest of brain as functions of numbers of neurons in these structures in primates (exponent, +1.2) and all other clades (exponent, +1.9). For comparable numbers of neurons in the rest of brain, these structures are much smaller in primates than in other animals.

primates. Because of the different scaling rules, a primate cortex, cerebellum, or entire rest of brain hid in it far larger numbers of neurons than expected for a nonprimate cortex, cerebellum, or rest of brain of comparable mass.

The structure- and order-specific neuronal scaling rules for the brain also told us much about the *nature* of the evolutionary changes underlying the divergences of the primate cortex and cerebellum, the eulipotyphlan cerebellum, and the primate rest of brain. It turned out we could make reasonable inferences about what must have happened when these mammalian clades emerged as new lineages. What made these inferences possible was, again, simple mathematical analysis.

Since the brain is made of cells, the size of a brain or brain structure is necessarily the result of how many cells it is made of and how large or small these cells are on average.* More exactly, the mass or volume of a brain or brain structure is the product of its number of cells—neurons and nonneuronal cells—and the average mass or volume of these cells. But, because we found the distribution of nonneuronal cells to be mathematically uniform across structures and species (more on that later), we can, for now, consider that whatever changes in the scaling rules that govern how brain structures are built are due mostly to changes in the way those structures gain neurons, rather than nonneuronal cells (glial and endothelial cells), of different sizes.

We knew about changes in numbers of neurons in mammalian evolution—but what could we say about the average size of these neurons? When the mass of a brain structure scales linearly with its number of neurons, as in the primate cerebral cortex and cerebellum, there is only one possible conclusion regarding the average size of the neurons: it must stay the same. To illustrate this, imagine a very elastic bean bag filled with varying numbers of Styrofoam beans. If only beads of a constant average size are used, then stuffing the bag with 3 times or 10 times more beads will make the bag exactly 3 or 10 times larger in volume. This is true even if not all beads are uniform in size, as long as the *average* size of all beads remains constant. In the case of brain structures, verifying whether the size of the beads (the neurons) is constant across species would be a daunting task because the volume of neurons is not concentrated locally: dendrites and especially axons extend sometimes very long distances away from the main body of the cell, and measuring the entire cell requires very labor-intensive three-dimensional reconstructions of microscopic sections of brain tissue.

But here is where simple mathematics comes yet again to the rescue. If the average size of neurons is indeed constant even as neurons vary in number, then the number of neurons per brain volume, that is, the density of neurons in the tissue, should also be constant, like the number of beads per unit of volume in the bag. Now this is very useful information because determining neuronal density is a trivial task with our method: we simply divide the number of neurons found in the tissue by the mass or volume of the tissue.

*In our calculations, the extracellular space between cells is included in the estimates of cell size, so there is no third component to the mass (or volume) of the brain besides number of cells and average cell size.

By the same token, filling the elastic bean bag with Styrofoam beads that *increase* in size as more beads are used results in a bean bag whose volume increases faster than it gains beads. If the bag is filled with 10 times more beads and each bead is now on average 1.5 times larger, then the final volume of the bag will be not just 10, but 10 × 1.5 = 15 times larger. If the beads are instead 5 times larger when added in numbers 10 times larger, then the bag becomes 5 × 10 = 50 times larger. Again, the change in average size of the beads in each bag can be inferred without measuring the actual size of the beads by simply determining, instead, the *density* of beads in the bag. In the first scenario, the density of beans in the bag beads is 1.5 times lower; in the second, it is 5 times lower. The density of the beads changes inversely with the change in average volume of the beads.* And what's more, if there is a constant relationship between the change in size of the beads as the beads become more numerous, this will appear as a fixed relationship between the density of beads and the number of beads in the bean bag. And if we substitute "neurons" for "beads" in these statements, we can learn how the average size of neurons changes as brain structures gain neurons in evolution.

The Primate Advantage

In the nonprimate cerebral cortex, as the number of neurons increases, neuronal density decreases in a uniform manner that can be described as a power function of the number of neurons with the allometric exponent –0.6 (figure 4.13). Because the average size of neurons varies inversely with neuronal density, this exponent implies that the average mass of nonprimate cortical neurons increases with the number of neurons in the cortex raised to the power of +0.6. This means that when a nonprimate cortex gains 10 times more neurons, its neurons become on average 4 times larger in mass, and the cortex thus 40 times larger in total mass. When it gains 100 times more neurons, its neurons become 16 times larger on average, and the cortex, 1,600 times larger; with 1,000 more neurons, they become 64 times

*The full demonstration that this applies to neurons in brain structures across mammalian species is given in Mota and Herculano-Houzel, 2014, where we showed that the calculated average mass of neuronal cells varies with measured neuronal density (neurons per milligram of brain tissue) raised to the power of –1.004, not significantly different from –1.000.

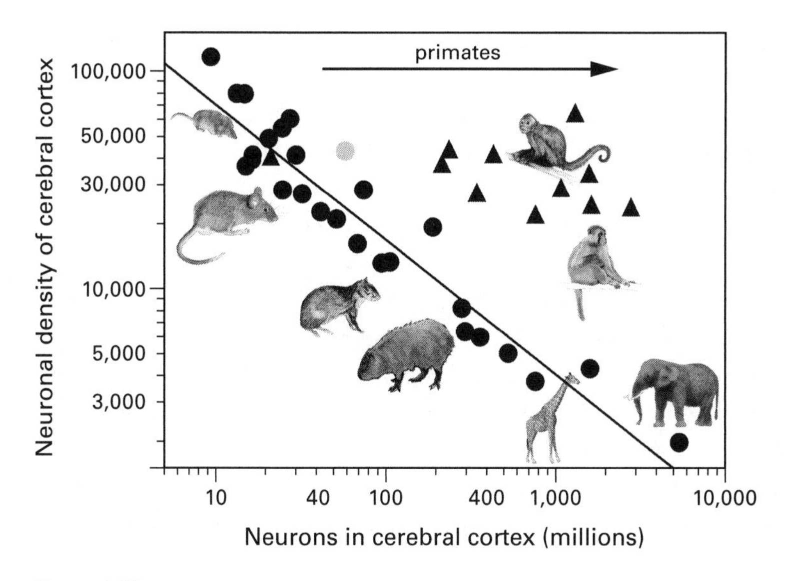

Figure 4.13
As the nonprimate cerebral cortex (circles) gains neurons, neuronal density (in neurons per milligram of cerebral cortex) drops with the number of neurons raised to the power of –0.6, which implies that the average mass of neurons increases with the number of neurons raised to the power of +0.6. In contrast, as the primate cortex (triangles) gains neurons, neuronal density does not decrease significantly. Considering that mammalian evolution started with small brains and cortices, it can be inferred that primates branched away from the common ancestor with changes that uncoupled the addition of neurons to increases in the average size of neurons in the cerebral cortex.

larger on average, and the cortex, a full 64,000 times larger. Across primate species, in contrast, as the number of cortical neurons increases, there is no systematic variation in neuronal density (figure 4.13). This means that, as the primate cortex evolved, it started to gain neurons without a significant systematic increase in the average size of its neurons. This does not mean that primate neurons became smaller, but only that they stopped growing: the primate cortex seems to have retained an average neuronal cell size similar to that in the rat or rabbit cortex, and definitely smaller than that in the agouti cortex. As a consequence, when a primate cortex gains 10 times as many neurons, it becomes only about 10 times as large, and if it gains 100 times as many neurons, it becomes only about 100 times as large.

The fairly tight correlation between neuronal density (and therefore average neuronal size) and number of neurons across nonprimates suggests that there must have been a mechanism that, in the evolution of nonprimate mammals, *coupled increases in the number* of neurons in the cerebral cortex *with increases in the average size* of these neurons. Every time that a yet-to-be-discovered mechanism increased the number of neurons building the cerebral cortex of a new nonprimate species, that mechanism also increased the average size of the neurons by a fixed amount. Moreover, this mechanism must have been (and continues to be) quite robust, for it has operated for more than 90 million years, since before the divergence of the afrotherian, rodent, eulipotyphlan, and artiodactyl lineages, and, indeed, it is still reflected in the modern species of each of these groups. When primates, for their part, branched off from the ancestor they had in common with nonprimates, it was with a break that included overriding this mechanism and *un*coupling increases in the number of cortical neurons from increases in the average size of the neurons. That is, whatever increased the number of cortical neurons from one primate species to the next, new, larger-brained species did *not* also make its cortical neurons any larger. And thus the neuronal scaling rules that apply to primates became different, and specific to them.

Because the first primate species were smaller than the current ones,[17] that "break" in the origin of primates would have appeared in animals with small numbers of neurons in their cerebral cortex, which is predicted to have had a number of neurons that matched both the nonprimate and the future primate scaling rules, that is, to have been located at the intersection of the primate and nonprimate distributions shown in figures 4.8 and 4.13. The current distribution of neuronal density values in modern primate species can thus be used to trace back this ancestral primate to the point where the nonprimate and primate functions converge in figure 4.13. The fossil of the stem primate *Ignacius graybullianus* suggests it had an endocranial volume of 2.14 cubic centimeters and a predicted body mass of 231 grams.[18] If this species was indeed positioned close to the branching off of primates, the ancestral scaling rules for mammalian brains must still have applied, in which case we can infer an approximate mass of 0.96 gram in the cerebral cortex, yielding an estimated 42.4 million neurons in it, close to the number of neurons found in the cerebral cortex of the modern mouse lemur.[19] The mouse lemur, by the way, sits just where the ancestral primate would be expected to be located in our graphs: at the intercept between primate and nonprimate scaling rules for the cortex. Using instead the scaling rules that we propose that applied to ancestral mammals, we find an intermediate value of 33.9 million neurons in the cerebral cortex. The convergence between the two estimates is what we would expect in our proposed scenario of branching of primates away from a common ancestor shared with modern nonprimates with modifications in neuronal scaling rules that became more and more noticeable as numbers of neurons increased across species in evolution.

What does it mean that the primate cerebral cortex gains neurons that do not on average become larger as they become more numerous, in contrast to other mammals, for which more neurons in the cortex necessarily mean larger neurons? First and foremost, it means that nature is not limited to one single way of building a cortex. But second, there is an important advantage to primates from not having neurons also become larger as they become more numerous: the cortex increases in volume by only as much as it gains neurons, and not more. In terms of volume, adding neurons that do not increase significantly in average mass is a very economical way to add neurons to the cortex, compared to the nonprimate alternative that rapidly leads to a cortex that is much larger for its number of neurons. This is important because volume comes at a premium: for example, the larger the cortex, the longer it will take for action potentials to propagate down fibers across the cortex, delaying the integration of information in the brain. Indeed, we have estimated that average propagation time for signals in the cortical white matter scales rapidly across rodents, as a power function of the number of cortical neurons raised to the power of +0.466, whereas, across primates, propagation time scales much more slowly as the cortex gains neurons with an allometric exponent of only +0.165.[20] The primate way of putting together a cerebral cortex brings the clear advantage of *reducing the increase in signal propagation time as the cortex gains neurons* (and thus increases in size), compared to nonprimates.

This raises an apparent conundrum, however: even if a primate cortex becomes only proportionately larger as it gains neurons, distances across the same two points in the cortex still necessarily increase and must therefore be bridged by longer, larger, neurons—and yet, neurons appear to retain their average size. How can both be true at the same time? The solution to the conundrum lies in the key concept of *average* neuronal size: it means that some neurons may very well become larger (longer, in this case), but if that happens, it is at the expense of others becoming shorter. In terms of the cerebral cortex and its underlying white matter, this means that if some cortical neurons in a larger brain now become larger, with longer fibers through the white matter, even more cortical neurons must now restrict their fibers to the gray matter. If neuronal size on average does not change significantly, or changes very little, across primate species, then the first type of neurons (which establish long-range connections through the subcortical white matter) must increase in numbers more slowly than the second type of neurons (which establish short-range connections within the gray matter, without ever crossing the border into the white matter). Indeed, in separate studies of how the subcortical white matter scales in volume as the cortex gains neurons, we could determine that cortical connectivity through the white matter—that is, the fraction of cortical neurons that are connected with long fibers through the white matter—does *decrease* as primate cortices gain neurons, as predicted in the scenario above. This is exactly how a small-world network becomes larger: not by scaling all connections equally, but by adding many more local connections and only a few, key, long connections. In contrast, cortical connectivity through the white matter remains stable as rodent cortices gain neurons and these neurons become, on average, larger.[21]

For the cerebellum, we find that neuronal density decreases as the cerebellum gains more neurons across nonprimate, noneulipotyphlan species, as shown in figure 4.14, indicating that average neuronal size increases jointly with numbers of cerebellar neurons in these species, according to a power function with the allometric exponent +0.3. This means that a tenfold increase in the number of neurons that build a nonprimate, noneulipotyphlan cerebellum is accompanied by neurons that become, on average, twice as large; a 1,000-fold increase in the number of these neurons is accompanied by an eightfold increase in the average size of cerebellar neurons. As a result, primate cerebellums, regardless of size, have neuronal densities—and thus average neuronal cell sizes—comparable to those of a rat cerebellum.

Remarkably, the increase in average neuronal size is much slower in the cerebellum than in the cerebral cortex as these structures gain neurons in similar numbers across nonprimate species. The discrepancy makes sense

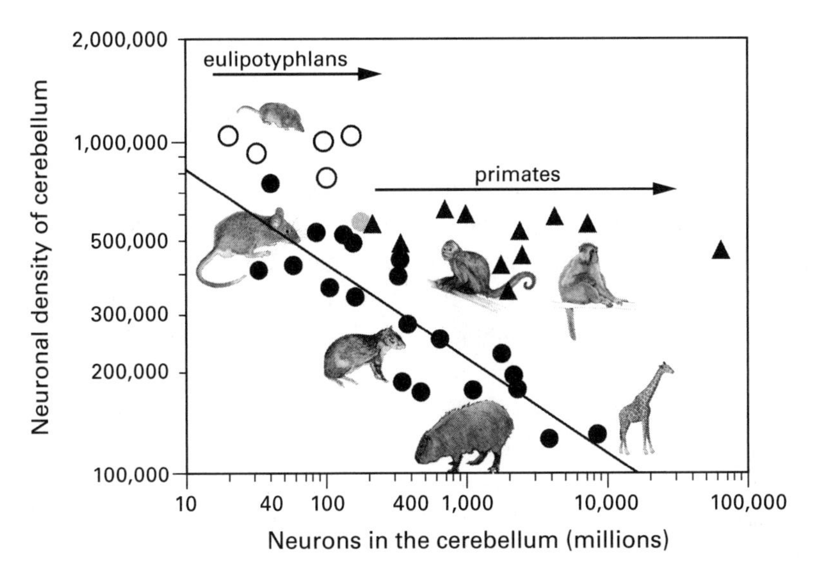

Figure 4.14
As the nonprimate, noneulipotyphlan cerebellum gains neurons (filled circles), neuronal density drops with the number of neurons raised to the power of −0.3, which implies that the average mass of neurons increases with the number of neurons raised to the power of +0.3. In contrast, as the cerebellum gains neurons across primate (triangles) and eulipotyphlan species (open circles), there is no significant decrease in neuronal density. Considering that mammalian evolution started with small brains and cortices, it can be inferred that primates and eulipotyphlans branched away from their respective ancestors with changes that uncoupled increases in the number of cerebellar neurons from increases in their average size.

with how the anatomy of the two structures compare. The reason behind the slower scaling of average neuronal size in the cerebellum than in the cortex is likely related to the difference in the pattern of long-range connectivity in the two structures. Whereas the cerebral cortex has a large number of long-range, reciprocal connections between cortical sites through the subcortical white matter, the cerebellar white matter consists uniquely of input or output connections *to* the cerebellar cortex, but not *across* cerebellar sites. There is, for instance, nothing comparable to a corpus callosum within the cerebellum, connecting the two halves. Instead, the longest fibers span only several millimeters (as opposed to centimeters in the cerebral cortex), and they do so within the upper layer of the cerebellar cortex itself, spread neatly in parallel along the cerebellar folia, where they distribute information along Purkinje cells. The arrangement of fibers in the cerebellum is thus geometrically more economical than that found in the cerebral cortex, and it is the basis of the many differences in local function of the two structures. Whereas the long-range fibers through the subcortical white matter allow columns of neurons in the cerebral cortex to process information that comes from centimeters away in convergent, divergent, and reciprocal patterns, the much more numerous parallel fibers of the upper layer of the cerebellum (for there is one for every single granule cell neuron) spread the information laterally over a few millimeters to be integrated locally by Purkinje cells.

Eulipotyphlans and primates seem to have independently broken away from their respective ancestral cerebellar scaling rules, each with modifications that uncoupled increasing numbers of neurons from increasing cell size, as illustrated in figure 4.14. These patterns imply that in the evolution of eulipotyphlans and primates, neurons began to be added to the cerebellum without any further significant increases in their average size. The lack of an increase in average neuronal cell size in the cerebellum of eulipotyphlans and primates suggests that those parallel fibers in the upper layer of the cerebellum still spread signals laterally over the same or shorter distances in large and small primate or eulipotyphlan cerebellums, unless the Purkinje cells in these larger cerebellums became smaller, in which case the parallel fibers spread signals laterally over longer distance. The latter possibility seems unlikely, though; in those assorted mammalian species where the size of Purkinje cells has been examined, it appears to increase with brain volume.[22] Still, the most important finding is that eulipotyphlans and primates add neurons to their cerebellums without the average neuron becoming any larger, which indicates that there is no one single way of building a larger cerebellum—just as there is no single way of building a larger cerebral cortex.

What It Means to Be a Primate

A few numbers show how dramatic are the consequences for primates of the break with the ancestral rules for putting together a cerebral cortex. As seen in figure 4.15, whereas the smallest primate is not too different in its number of cortical neurons from a nonprimate of similar cortical mass,

Nonprimates			Primates		
398.8 g	Giraffe	1.7 B	~ 6 B	Chimpanzee	286.0 g
111.3 g	Blesbok	571 M	2.9 B	Baboon	120.2 g
68.8 g	Springbok	397 M	1.7 B	Rhesus monkey	69.8 g
42.2 g	Pig	303 M	1.6 B	Bonnet monkey	48.3 g
8.9 g	Agouti	111 M	442 M	Owl monkey	10.6 g
4.4 g	Rabbit	71 M	245 M	Marmoset	5.6 g
0.9 g	Spiny rat	26 M	22 M	Mouse lemur	0.9 g

Figure 4.15
Cortical mass (in grams) and number of neurons (in millions, M, or billions, B) in different nonprimate and primate species, according to our estimates.

the larger the cerebral cortex of a primate becomes, the more extreme becomes its advantage over other animals in sheer numbers of neurons. Similar numerical advantages apply to the primate cerebellum and rest of brain over other nonprimate mammalian species. Primates have a clear advantage over other mammals, which lies in an evolutionary turn of events that resulted in the economical way in which neurons are added to their brain, without the massive increases in average cell size seen in other mammals.

We had an answer to our initial question, then: all brains were *not* made the same. In particular, primate brains were *not* made the same way as nonprimate brains. Although a cow and a chimpanzee have brains of about the same mass, the chimpanzee can be expected to have at least twice as many neurons as a cow. Actually, even the brains of nonprimate species are put together in different ways among individual groups. Although the neuronal scaling rules shown here indicate that rodents, eulipotyphlans, afrotherians, and artiodactyls share the same relationship between cerebral cortical volume and number of neurons, we later discovered that the volume of neurons is spread out more thinly or thickly in the cerebral cortex in each different group, just like a similar spoonful of jam can be spread more thinly or thickly over small or large pieces of toast—but that is a whole other story. For now, the burning question is, what about *us*?

5 Remarkable, but Not Extraordinary

Comparing the cellular composition of the brain of a large number of a mammalian species showed that not all brains were made the same. Two brains of similar size did not necessarily share similar numbers of neurons, and a larger brain did not necessarily have more neurons than a smaller one. That is, not if only one of them belonged to a primate, for the primate way of putting together a cerebral cortex and cerebellum was more economical in terms of volume, with a break away from the otherwise seemingly mandatory coupling of more neurons with larger neurons: because primate neurons in the cerebral cortex and cerebellum stopped becoming larger as they became more numerous, primate brains are made with far more neurons in these two structures than one would suspect from the size of the brain. We could now address the burning question of how the human brain compared to others.

Once we knew the neuronal scaling rules for rodent brains, in 2006, we could already do some rough calculations. With the equations relating the number of neurons in a rodent brain to the mass of the brain and of the body, we could estimate that a rodent brain that had anything in the order of 100 billion neurons, as the human brain was supposed to have, would weigh more than 30 kilograms and belong in a body that weighed more than 80 tons.[1] In other words: if we were rodents, we would look like a blue whale, have to live in water, and carry an impossibly large brain, one that would likely collapse under its own weight. Compared to that, the fact that we can carry our weight on two flimsy legs walking on land and balance a very modestly sized brain in our heads makes us look extraordinary indeed.

But comparing humans to something we most obviously are not and then, based on the preposterous outcome, deducing that we are out of the ordinary is just not science. This comparison only shows what we already knew: that we are not rodents. We don't have huge incisor teeth; we don't have claws or eyes facing sideways. And even though we belong to a group of animals that are closer to rodents than to other mammalian groups, our closest relatives in our own group are those equally equipped with binocular vision, five-fingered hands with fingernails, and dexterous fingers that move independently: we are primates.

The proper comparison to determine whether the human brain was at all out of the ordinary was therefore to compare it to the brain of fellow primates. It was 2007, and we already knew the neuronal scaling rules that applied to primate brains. The question that remained to be answered, then, was, given the average number of neurons in the human brain, did it have just the size predicted for a generic primate brain of its number of neurons, or was it special—extraordinarily small, or large?

If only we knew how many neurons a human brain was made of.

Dissolving Human Brains

In our work with the Pathology Department at the Federal University of Rio to figure out what human brains were made of, we were having technical difficulties. The brains we had been able to collect had been strongly overfixed, having sat in buckets of formaldehyde for months until they could be examined, and the overfixation had led to a high concentration of aldehydes in the tissue that rendered all nuclei strongly fluorescent in both green and red before we even had a chance to use our stains to determine which cell nuclei belonged to neurons. I tried a number of protocols that looked very much like cookbook recipes to get rid of the excess fluorescent aldehydes: I stewed the nuclei in citric acid, cooked them in the microwave, washed them again and again in different solutions. I tried bleaching them under colored lights. Nothing worked. As I peered through the microscope, the nuclei still stared back at me in all their bright fluorescence, precluding any attempts to use stained antibodies to determine how many of them were neurons.

The solution was a nonsolution: we acknowledged the impossibility of using those overfixed brains and embraced the opportunity to start a new collaboration with the team led by Léa Grinberg and Wilson Jacob Filho at the School of Medicine of the University of São Paulo (FMUSP), which runs that city's death certification service and an associated brain bank, organized as part of the Center for the Study of Aging. By then, Frederico Azevedo, a Master's student in Roberto Lent's lab, had asked to be assigned to the human brain project, and I became his coadvisor with Roberto. Fred began to work in conjunction with Léa, Renata Leite, Renata Ferretti-Rebustini, José Marcelo Farfel, and Wilson Jacob Filho, the team at FMUSP, to ensure that they would be able to fix the donated brains only very lightly, but thoroughly by perfusing them with paraformaldehyde through the carotid arteries instead of simply dropping them into a bucket of fixative (the standard procedure in many brain banks). Fred found that if, after perfusing the brains, he immersed them in fixative for just a few days, he could turn them into soup and get nuclei that were still perfectly intact and reactive to the anti-NeuN antibody—but that were not yet autofluorescent.

We were in business.

Four brains and a year later, we had our results (and Fred had his Master's thesis). I should point out that our brain data were average numbers for (1) Brazilian (2) males (3) aged 50 or 70 years—so there was nothing we could say at this point about male-female differences, individual variations, aging effects, or differences across ethnicities. Yet, for the purposes of comparing averages across human and nonhuman primate species that range in brain mass, body mass, and number of brain neurons by several orders of magnitude, our averages would do just fine.

And those averages were 16 billion neurons in the human cerebral cortex, 69 billion in the cerebellum, and slightly fewer than 1 billion in the rest of brain, for a total of 86 billion neurons in the entire human brain. For those who like to observe that "86 is close to 100" and who claim that the original order-of-magnitude approximation is therefore accurate (and, granted, as an order-of-magnitude estimate, it is), I like to point out that the missing 14 billion neurons represent an *entire baboon brain*—with 3 billion neurons to spare. There was variation across our individual donors, of course—and yet not one came within striking distance of the mythical 100 billion neurons in his brain, although our oldest brain did have 91 billion neurons (a "mere" 9 billion away).

It is amazing how much can be learned from just these numbers, as I will point out in the chapters to come. For now, however, the most important thing about the 86 billion neurons is where they place humans in comparison to other primates. According to the neuronal scaling rules that apply to primates, we could expect a generic primate brain with a total of 86 billion neurons to weigh about 1,240 grams (2.75 pounds) in a body weighing about 66 kilograms (145 pounds). These numbers are just about right for us humans, with our, on average, 1,500-gram (3.3-pound) brains and 70-kilogram (155-pound) bodies. The conclusion should come as no surprise to a biologist: we are that generic primate with 86 billion neurons in its brain. Our brain is made in the image of other primate brains. In comparison to a generic rodent with a similar number of neurons in its brain (figure 5.1), we do fit all those neurons in an extraordinarily small volume. But we are not rodents, we are primates—and with a perfectly ordinary primate brain. Well, at least in terms of total number of brain neurons.

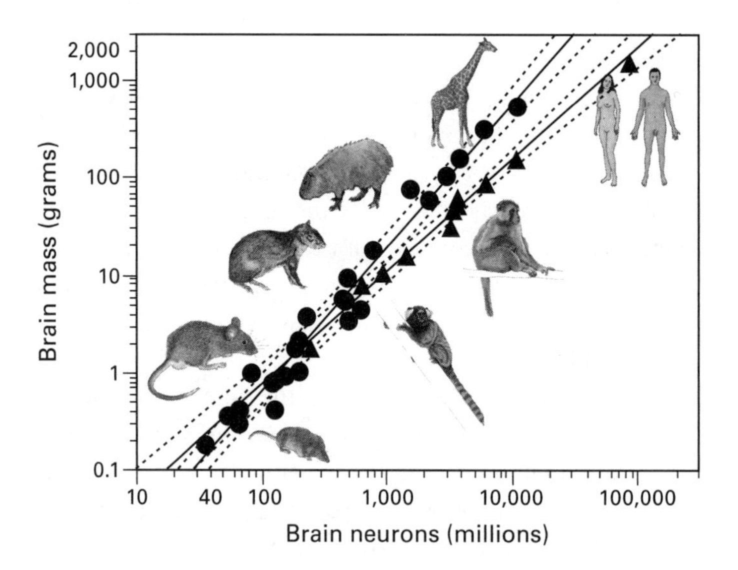

Figure 5.1
The human brain has the relationship between brain size (mass) and number of neurons that would be expected for a generic primate. The power function plotted for nonprimates (circles) has an exponent of 1.5, whereas the power function plotted for primates (triangles), excluding the human species, has an exponent of 1.1. The dashed lines indicate the 95 percent confidence intervals for each function—and the fact that the human species is well contained within that interval for primates indicates that it obeys the same neuronal scaling rule that applies to other primate species.

In the Image of Other Primate Brains

Again, it could be the case that the human brain only *appeared* to conform to the neuronal scaling rules that applied to other primates, but in fact did not—for instance, if a much larger than expected number of those neurons were in the cerebral cortex (which would support the claims that our cortex is overexpanded), compensated for by a much smaller than expected number of neurons in the cerebellum. That was easy to check, since we had counted the neurons for the cerebral cortex, cerebellum, and rest of brain separately.

We found that the total number of neurons in the human cerebral cortex, 16 billion neurons on average, is close to (and even slightly smaller than) the 19.9 billion neurons expected for a generic primate cortex with its cortical mass, 1,233 grams. As shown in figure 5.2, our cortex falls within the 95 percent confidence interval used routinely to test for conformity. The human cerebral cortex is therefore not outstanding in its neuronal composition: it has just the mass that a primate cortex with its number of neurons would be expected to have.

The same was found for the human cerebellum. At an average 154 grams and 69 billion neurons, it also falls near the expected combination for a generic primate (figure 5.3). The human cerebellum is therefore also not extraordinary in its neuronal composition as a primate cerebellum. Nor is the human rest of brain special in its composition: as seen in figure 5.4, it shares the same relationship between structure mass and number of neurons that applies to other primates.

Our findings meant, among other things, that whatever had happened in the evolutionary history of humans that led to our very enlarged brain compared to other primates (we did, after all, gain a three times bigger brain compared to the last ancestor we shared with our nearest living cousins, chimpanzees and bonobos, and in the record time of just 1.5 million years to boot), it was with a brain that still obeyed the same neuronal scaling rules that had applied to the primates before us. Our brain is made in the image of other primate brains. The thought that Darwin would have appreciated our findings puts a big smile on my face.

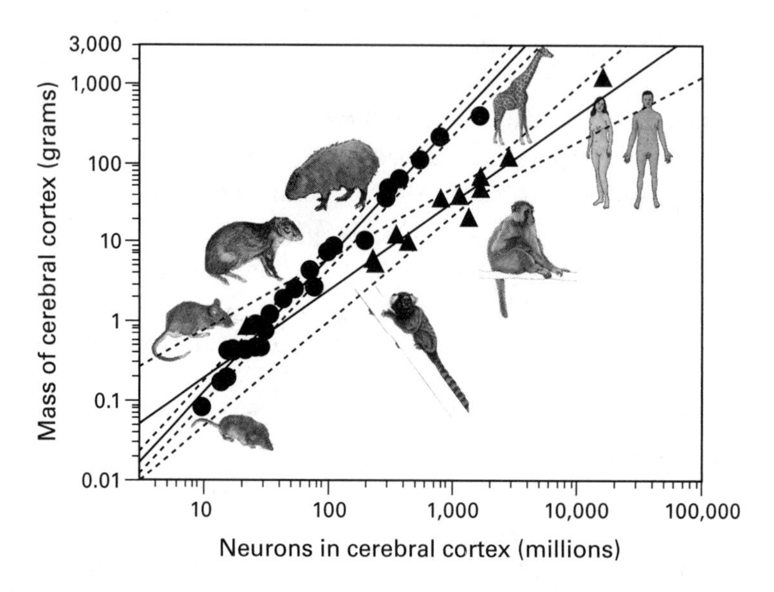

Figure 5.2
The human cerebral cortex has the mass expected for a generic primate with its number of neurons (or the number of neurons expected for its mass). The power function plotted for nonprimates (circles) has an exponent of 1.6, whereas the power function plotted for primates (triangles), excluding the human species, has an exponent of 1.1. The dashed lines indicate the 95 percent confidence intervals for each function—and the fact that the human species is well contained within that interval for primates indicates that its cerebral cortex is made according to the same neuronal scaling rule that applies to the cortex of other primate species.

Human Evolution: Great Apes and Hominins

We had what we thought was a very straightforward set of findings, which should have been just as straightforward to publish and make widely available to our fellow scientists and the lay public alike. We had, for the first time, a complete estimate of the average number of neurons and non-neuronal cells in the whole human brain, which was *not* 100 billion—and the average total number of glial cells was nowhere near 10 times more than the number of neurons, by the way, as we'll see later. The actual number of neurons was almost precisely what would be expected of a generic primate brain. "The human brain is just a scaled-up primate brain: remarkable, but not special" was our main message—and because we believed it was a big one, we set our sights on the star journals in the field.

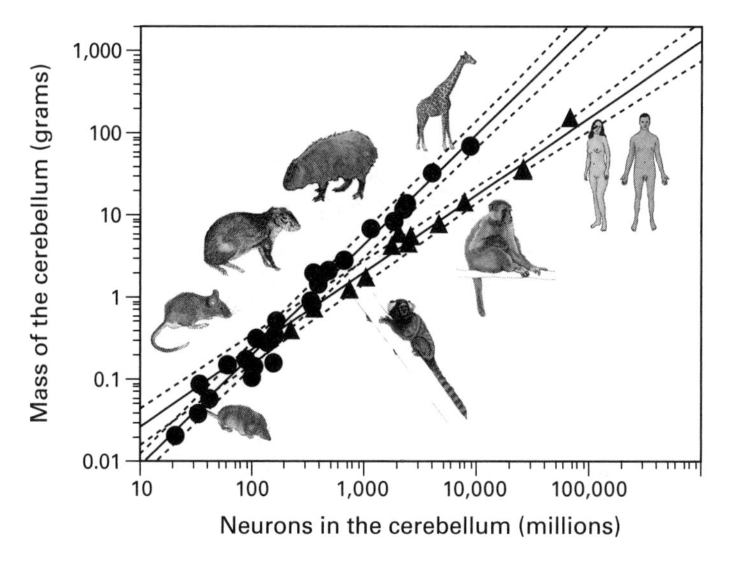

Figure 5.3
The human cerebellum has the mass expected for a generic primate with its number of neurons (or the number of neurons expected for its mass). The power function plotted for nonprimates (circles) has an exponent of 1.3, whereas the power function plotted for primates (triangles), excluding the human species, has an exponent of 1.0. The dashed lines indicate the 95 percent confidence intervals for each function—and the fact that the human species is well contained within that interval for primates indicates that its cerebellum is made according to the same neuronal scaling rule that applies to the cerebellum of other primate species.

And were rejected, again and again, for a variety of reasons. In retrospect, it amuses me to no end that "Equal Numbers of Neuronal and Non-Neuronal Cells Make the Human Brain an Isometrically Scaled-Up Primate Brain,"[2] which has become a highly cited paper (with more than 300 citations in other scientific papers in just five years), and which is commonly mentioned now in the opening lines of many papers on the human brain, was rejected without review by *Nature*, *Proceedings of the National Academy of Sciences of the U.S.A.*, *Neuron*, and the *Journal of Neuroscience*—journals that rank among the highest in neuroscience exactly because of how often their papers are cited in the years after the publication. We were finally accepted by the *Journal of Comparative Neurology*, after a long back-and-forth exchange with their reviewers, and, five years after publication, our 2009 paper still remained the one most visited

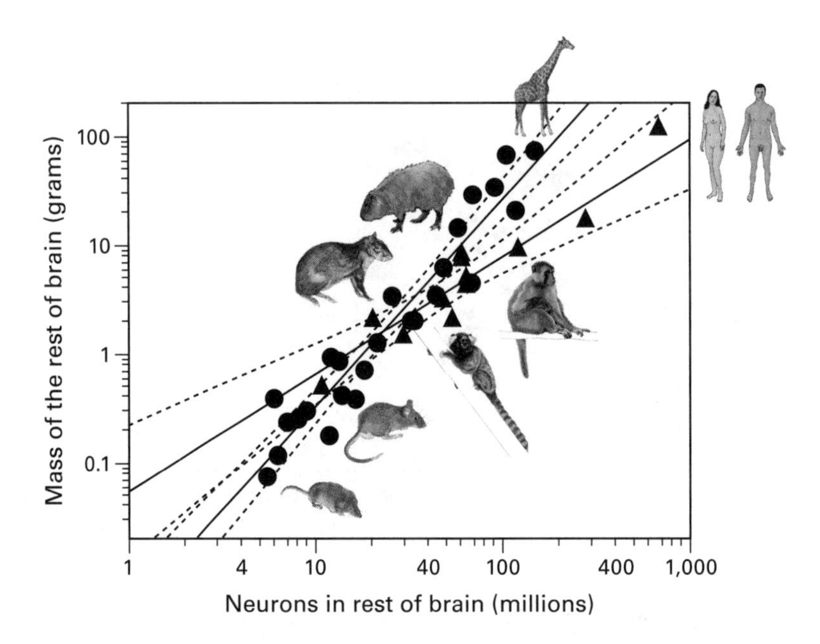

Figure 5.4
Human rest of brain has the mass expected for a generic primate with its number of neurons (or the number of neurons expected for its mass). The power function plotted for nonprimate species (filled circles) has an exponent of 1.9, which excludes primates (triangles), whereas the power function plotted for primates, excluding the human species, has an exponent of 1.2. The dashed lines indicate the 95 percent confidence intervals for each function—and the fact that the human species is well contained within that interval for primates indicates that its rest of brain is made according to the same neuronal scaling rule that applies to the rest of brain of other primate species.

on the journal's website. *Science* had sent the manuscript out to review, but rejected it after one reviewer claimed that our findings were "just what stereology had already shown," and thus "not novel," while another, who clearly misread our numbers, argued that they were too *different* from those of previous stereological studies, and thus not believable. During the first years of our work, we met often with this type of incredulous criticism: because stereology was a well-established method, we should verify our numbers against stereology first—a criticism that failed to grasp that the whole point of our method was exactly to go where stereology was *unable* to go within a reasonable time frame: the *whole* brain.

But there was another issue. We also had scaling rules for the number of brain neurons as a function of the mass of the body (more on this later), and we pointed out that the human brain had just as many neurons as would be expected for a non–great ape primate of its *body mass*—which flew in the face of Harry Jerison's assertion that the human brain is too large for its body. One of the reviewers for *Science* could just not accept that: given that gorillas and orangutans have larger bodies and yet smaller brains than humans, how could we claim that our brain was just the size it should be?

We had suggested how in the text, but the reviewer had disregarded that suggestion. The reason why neuroscientists had thought all along that the human brain was extraordinarily large for the body that housed it was exactly the direct comparison to great apes: if our body is smaller than the body of a gorilla, then our brain should also be smaller, and yet it is three times larger in mass. But our data showed that, when great apes were excluded (simply for lack of numbers of brain neurons for these species at that point), humans shared the same relationship between body mass and number of brain neurons as other primates. So, what if—just what *if*—instead of humans having brains that were too *large* for their bodies, it turned out that gorillas and orangutans had brains that were too *small* for their bodies?

To address that question, the first step was to figure out if great ape brains also conformed to the primate neuronal scaling rules. That required having great ape brains to analyze—a commodity that is understandably in very short supply (and I'm actually happy that's the case). But we did have four great ape cerebellums in storage, retrieved from Jon Kaas's cold storage room: one from a gorilla and three from orangutans. These were in very good condition, but hopelessly overfixed, so that using the staining for NeuN to quantify neurons was out of the question. But, as it happens, far and away most neurons in the cerebellum are small granule cell neurons, whose uniformly small and round cell nuclei are easy to distinguish from other nuclei. And even if we only had the cerebellums to work with, that would still be useful. As we'll see in greater depth in chapter 7, the scaling rules across brain structures are so tight that the number of cells in the cerebellum alone would already allow us to predict the size of the entire brain of a primate. If we could do that, then we could answer the important question of whether the brain of these great apes was also built according to the same scaling rules that applied to human and nonhuman primates alike. And if great ape brains *did* conform to the generic primate rules, then it followed that it was their size relative to their bodies that did not.

We found that the gorilla cerebellum contained 29 billion cells and the orangutan cerebellum 28 billion, 26 billion of which, in either case, were neurons.[3] Given their respective cerebellar masses of 38 grams and 35 grams, those numbers of cerebellar cells, compared to the expected 25 and 24 billion, placed them squarely in the scaling relationship that applied for other primates: the cerebellums of these great apes were standard, generic primate cerebellums in their cellular composition. Most important, the 29 and 28 billion cerebellar cells predicted total brain masses of 483 grams and 470 grams in the gorilla and orangutan, within 10 percent of the average 486 grams and 512 grams reported in the literature. Just as a reference, the total number of cells in the human cerebellum, 85 billion, predicts a human brain of 1,433 grams—whereas the actual average mass of the human brains we used to generate the scaling relationships applied to the great apes was a very close 1,509 grams. The fact that brain mass could be predicted so accurately for the gorilla and orangutan simply from the number of cells in their cerebellums had one clear meaning: they, too, possessed typical, generic primate brains—just as humans did.

Finding that the same scaling rules applied to the brains of humans, great apes, monkeys, and other simians also had a highly significant implication for investigating our evolutionary origins. Our species appeared less than 1 million years ago; our last shared ancestor with gorillas and orangutans lived some 16 million years ago; all primates share more than 50 million years of evolutionary history. If the same neuronal scaling rules that applied 50 million years ago still applied 16 million years ago (when the gorilla and orangutan lineages appeared) and less than 1 million years ago (when we appeared), that meant that they also applied to the species that lived in between: our probable hominin ancestors, like the australopithecines,who lived 4–3 milllion years ago, and *Homo erectus*, who lived 2–1 million years ago.

If the same rules applied, that meant that we could use fossil data on cranial capacity to infer the number of neurons in the brain of extinct hominin species, our own ancestors and others'. Using published data on the brain mass of these species predicted from cranial capacity,[4] we estimated (1) that, 6–7 million years ago, *Sahelanthropus tchadensis*, the most recent presumed common ancestor to living humans and chimpanzees, with a brain mass of 363 grams, had 25 billion brain neurons, some 7 billion of which in its cerebral cortex; (2) that, around 4 million years ago, australopithecines such as the bipedal-walking Lucy had between 30 and 34 billion brain neurons, some 9 billion of which in their cerebral cortex, similar to the number in the cortex of modern great apes; (3) that, 2 million years ago, the early *Homo* species *H. habilis, H. ergaster*, and *H. rudolfensis* had 40–50 billion brain neurons, 11 to 14 billion of which in their cerebral cortex; and (4) that, from 1.5 million years ago onward, *Homo erectus* underwent a jump to 50–60 billion neurons, 17 billion of which in the cerebral cortex. The same scaling rules (based only on our initial six primate species) predict 85–88 billion neurons in the brain of *Homo neanderthalensis* and *Homo sapiens*, 23–24 billion of which in their cerebral cortex alone. In contrast, from what we found in the cerebellums of gorillas and orangutans, we predicted not more than 33 and 32 billion neurons in their brains, 8–9 billion of which located in the cerebral cortex. The increase in the total number of brain neurons, predicted from the known cranial capacity of living and extinct primate species, is illustrated in figure 5.5.

Figure 5.5 also shows the predicted relationship between the number of brain neurons and body mass for living and extinct primate species, including hominins. The lower plotted function is the usual way of looking at this relationship: when only living species are considered, including great apes but excluding humans, it appears that *Homo sapiens* has at least three times as many neurons as would be expected for its body mass. That would have matched Jerison's view: we have way too many neurons, many more than we should have. We are outliers that escape the evolutionary rules that apply to other primates.

But if only hominins are considered, both living (us) and extinct, then we get a completely different picture—the upper plotted function in figure 5.5: a relationship between body mass and number of brain neurons that covers not only living humans, their direct ancestors and other closer relatives, but also most other living primate species—and that excludes living great apes, with their far too few brain neurons for their body mass. In this

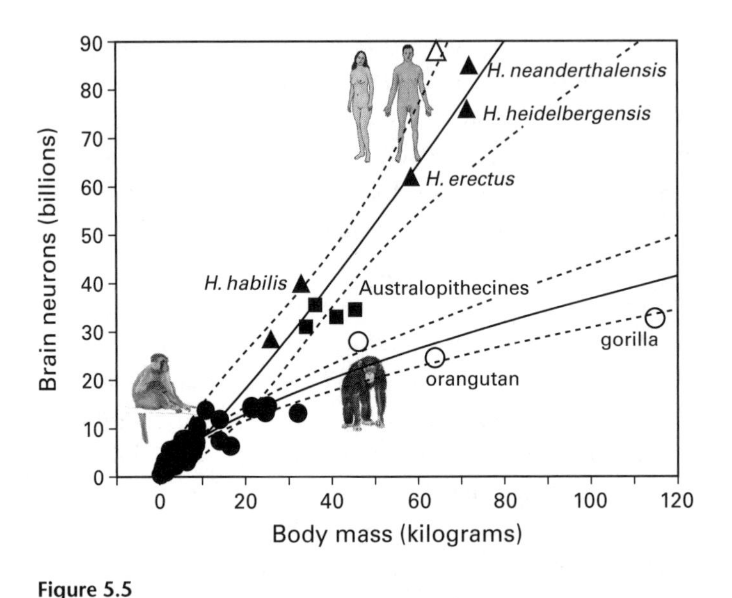

Figure 5.5
Numbers of brain neurons predicted in living and extinct primate species from the neuronal scaling rules that apply to nonhuman, non–great ape primates. *Sahelanthropus tchadensis* (not shown, for its body mass is unknown) is predicted to have had slightly fewer neurons than living great apes. The upper function is plotted exclusively for *Homo* (triangles) and its australopithecine ancestors (squares), but it predicts the number of brain neurons and body mass for most non-great ape primates; the lower function is plotted exclusively for living nonhuman primate species, including great apes (open circles), and it excludes humans. Thus, whether or not modern humans fit the scaling rules for other modern primates depends on whether or not great apes are included in the comparison—a clear sign that great apes might be the outliers themselves.

scenario, it is not humans that are outliers in their brain-body relationship; it is great apes. We are not special.

Such was the evidence we had at that point. Our numbers showed that the great ape brain was just another primate brain, like ours. Our brain, in turn, conformed in size and number of neurons to the body mass of other, non–great ape primates—but theirs did not. Wasn't it possible, then, that the great apes, for some reason, didn't fit the rules? That *they* had broken away from other primates in the relationship between number of brain neurons and body mass? It may sound like empty semantics (are you smaller than I or am I larger than you?), but the distinction is enormously important in terms of evolution, for it establishes what diverged away from the standard blueprint: was it our brains that became too large, or theirs that became relatively small? Had brain mass somehow become dissociated from body mass in great apes? And, even more important, how come?

I had a suspicion, based on what I was just beginning to learn about the scaling of the metabolic cost of the brain as a function of its number of neurons, that the reason large great apes had very large bodies but not the large brains that should go with them was a metabolic limitation—they simply couldn't afford both. This suspicion would lead to one of our most important realizations, which addressed the uniqueness of humans and how we came to exist at all with our distinctively large number of brain neurons—but without ever drifting away from the primate way of putting the brain together.

But before we go there, there is an equally important issue we need to address first: does this remarkable, but not extraordinary, number of neurons in the human brain really provide a basis for our outstanding cognitive abilities?

6 The Elephant in the Room

We have long deemed ourselves to be at the pinnacle of cognitive abilities among animals. But that is different from being at the pinnacle of evolution in a number of very important ways. As Mark Twain pointed out in 1903,[1] to presume that evolution has been a long path leading to humans as its crowning achievement is just as preposterous as presuming that the whole purpose of building the Eiffel Tower was to put that final coat of paint on its tip. Moreover, evolution is not synonymous with progress, but simply change over time. And humans aren't even the youngest, most recently evolved species. For example, more than 500 new species of cichlid fish in Lake Victoria, the youngest of the great African lakes, have appeared since it filled with water some 14,500 years ago.[2]

Still, there is something unique about our brain that makes it cognitively able to ponder even its own constitution and the reasons for its own presumption that it reigns over all other brains. If we are the ones putting other animals under the microscope, and not the other way around,* then the human brain must have something that no other brain has.

Sheer mass would be the obvious candidate: if the brain is what generates conscious cognition, having more brain should only mean more cognitive abilities. But here the elephant in the room is, well, the elephant—along with a whole series of cetacean species that are also larger brained than humans, but not equipped with behaviors as complex and flexible as ours. Besides, equating larger brain size with greater cognitive capabilities presupposes that all brains are made the same way, starting with a similar relationship between brain size and number of neurons,

*Amusing science fiction stories notwithstanding, like the mice in Douglas Adams's universe who have been studying human scientists all along...

and we have just seen that primate brains are made differently from other brains.

Now that my colleagues and I knew how many neurons different brains were made of, we could rephrase "more brain" and test it. Sheer *number of neurons* would be the obvious candidate, regardless of brain size, because if neurons are what generates conscious cognition, then having more neurons should mean more cognitive capabilities. Indeed, even though cognitive differences among species were once thought to be qualitative, with a number of cognitive capabilities once believed to be exclusive to humans, it is now recognized that the cognitive differences between humans and other animals are a matter of degree. That is, they are *quantitative*, not qualitative, differences. Our tool use is impressively complex, and we even design tools to make other tools—but chimpanzees use twigs as tools to dig for termites, monkeys learn to use rakes to reach for food that is out of sight,[3] and crows not only shape wires to use as tools to get food, but also keep them safe for later reuse.[4] Alex, the African gray parrot owned by psychologist Irene Pepperberg, learned to produce words that symbolize objects,[5] and chimpanzees and gorillas, though they cannot vocalize for anatomical reasons, learn to communicate with sign language.[6] Chimpanzees can learn hierarchical sequences: they play games where they must touch squares in the ascending order of the numbers previously shown, and they do it as well and as fast as highly trained humans.[7] Chimpanzees and elephants cooperate to secure food that is distant and can't be reached by their efforts alone.[8] Chimpanzees, but also other primates, appear to infer others' mental state, a requirement for showing deceitful behavior.[9] Even birds seem to have knowledge of other individuals' mental state, as magpies will overtly cache food in the presence of onlookers and then retrieve and move it to a secret location as soon as the onlookers are gone.[10] Chimpanzees and gorillas, elephants, dolphins, and also magpies appear to recognize themselves in the mirror, using it to inspect a visible mark placed on their heads.[11]

These are fundamental discoveries that attest to the cognitive capacities of nonhuman species—but such one-of-a-kind observations do not serve the types of cross-species comparisons we need to make if we are to find out what it is about the brain that allows some species to achieve cognitive feats that are outside the reach of others. And here we run into another problem, the biggest one at this point: how to measure cognitive capabilities in a large number of species and in a way that generates measurements that are comparable across all those species? Numbers don't solve every problem, of course, but sometimes they're quite useful. In this particular case, what we needed was a way to express the cognitive capabilities of a species in numbers that could be compared to other species.

Exactly because cognitive capabilities are multiple, one way to achieve valid cross-species comparisons of such capabilities has been to multiplex many of them tested independently and to concoct a single number that represents an index of "global cognition," which can then be compared across species. This meta-analysis approach, which is necessarily limited to species that are fairly close to one another in body shape, habitat, and interests (because a viable task for a primate with nimble fingers and a taste for candied fruits that fit inside small holes will not apply to, say, a dog or an elephant), was used by Robert Deaner and colleagues in 2007 to generate indices of global cognition for nonhuman primate species. Interestingly, what the authors found[12] was that the brain-related parameters whose variation best aligned with variation in the global cognition index across species were absolute brain mass and cortical mass—not the encephalization quotient, not relative cortical mass, and not residual variations in brain or cortical mass beyond what could be expected from body mass. In other words: the larger the brain and the larger the cerebral cortex of a nonhuman primate (since the two are nondissociable across primate species), the more it is able to achieve in terms of cognition. Because we now know that larger primate brains are made of proportionately larger numbers of neurons, it then follows that sheer absolute number of neurons is a better proxy for cognitive abilities than the encephalization quotient, at least across nonhuman species.

A similar finding, this time extendable across a wider range of species, was made more recently by a multinational team led by Evan MacLean in 2014.[13] Using the opposite strategy of restricting the tests to just two, precisely replicated across species, researchers from twelve countries applied the same two tests to the subjects of their expertise—mostly primates, but also small rodents, doglike carnivores, the Asian elephant, and a variety of bird species. The tests involved self-control, a cognitive ability that relies on the prefrontal, associative part of the cerebral cortex. One test, for example, gauged the ability of the subject species to refrain from seeking food in place A, where it had been visibly hidden and successfully retrieved before, upon seeing it moved to a new location, B. For human bystanders watching the task, it is obvious that the food must now be retrieved from B, not A—but that is exactly because we have a human brain and other animals don't. Most species required a certain number of attempts until they started to aim for location B instead of persevering on A. And, again, the best correlate with correct performance in the test of self-control was absolute brain volume—except for the African elephant, which, despite being the largest brained in the set, failed miserably at the task. A number of reasons come to mind, from "It did not care about the food or the task" to "It enjoyed annoying its caretakers by not performing." (I like to think that the reason why it's so hard to train monkeys to do things that are easily learned by humans is their dismay at the obviousness of the task: "C'mon, you want me to move to do just *that*? Gimme something more challenging to do! Gimme videogames!")

The most interesting possibility to me, however, is that the African elephant might not have all the prefrontal neurons in the cerebral cortex that it takes to solve self-control decision tasks like the ones in the study. Once we had recognized that primate and rodent brains are made differently, with different numbers of neurons for their size, we had predicted that the African elephant brain might have as few as 3 billion neurons in the cerebral cortex and 21 billion neurons in the cerebellum, compared to our 16 billion and 69 billion, despite its much larger size—if it was built like a rodent brain. On the other hand, if it was built like a primate brain, then the African elephant brain might have a whopping 62 billion neurons in the cerebral cortex and 159 billion neurons in the cerebellum. But elephants are neither rodents nor primates, of course; they belong to the superorder Afrotheria, as do a number of small animals like the elephant shrew and the golden mole we had already studied—and determined that their brains did, in fact, scale very much like rodent brains, as seen in chapter 4.

Here was a very important test, then: did the African elephant brain (figure 6.1), more than three times as heavy as ours, really have more neurons than our brain? If it did, then my hypothesis that cognitive powers come with sheer absolute numbers of neurons would be disproved. But if the human brain still had many more neurons than the much larger African elephant brain, then that would support my hypothesis that the simplest explanation for the remarkable cognitive abilities of the human species is the remarkable number of its brain neurons, equaled by none

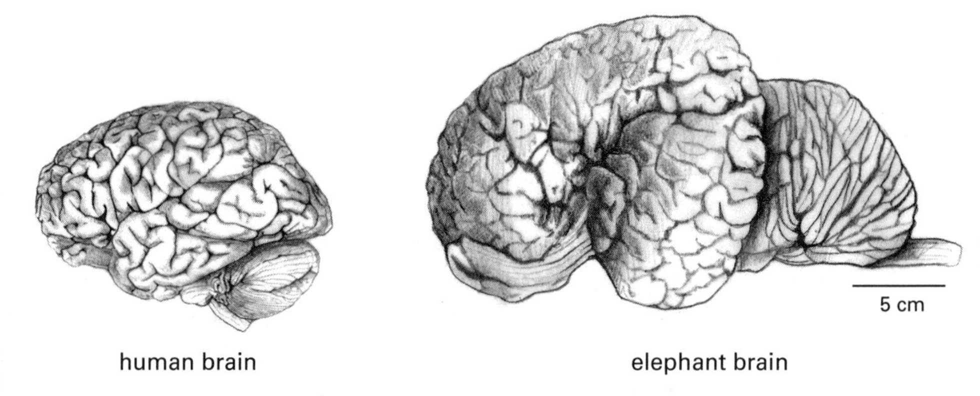

Figure 6.1
Side-by-side view of human and African elephant brains.

other, regardless of the size of the brain. In particular, I expected the number of neurons to be larger in the human than in the African elephant cerebral cortex.

The logic behind my expectation was the cognitive literature that had long hailed the cerebral cortex (or, more precisely, the prefrontal part of the cerebral cortex) as the sole seat of higher cognition—abstract reasoning, complex decision making, and planning for the future. However, nearly all of the cerebral cortex is connected to the cerebellum through loops that tie cortical and cerebellar information processing to each another, and more and more studies have been implicating the cerebellum in the cognitive functions of the cerebral cortex,[14] with the two structures working in tandem. The view that acknowledges the role of the cerebellum as a modulator of cortical processing as a whole has been slowly replacing the more usual view of the cerebellum as a structure merely necessary for sensorimotor learning. Because we had already found that the cerebral cortex and cerebellum gained neurons in concert in mammalian evolution (more on this in chapter 7), cognitive capabilities should scale not just with the number of neurons in the cerebral cortex alone, but with both the cortex and cerebellum together as two nondissociable structures. And, because these two structures together accounted for the vast majority of all neurons in the brain, cognitive capabilities should correlate equally well with the number of neurons in the whole brain, in the cerebral cortex, and in the cerebellum.

Which is why our findings for the African elephant brain turned out to be better than expected.

Brain Soup by the Gallon

The brain hemisphere of an African elephant that Paul Manger had made available weighed more than 2.5 kilograms, which meant that it would obviously have to be cut into hundreds of smaller pieces for processing and counting since turning brains into soup works with chunks of no more than 3–5 grams of tissue at a time. I wanted the cutting to be systematic, instead of haphazard: if we had a complete series of coronal cuts of the elephant brain, like a loaf of bread sliced in its entirety, and if we then cut each of those brain slices systematically into smaller pieces, following natural landmarks (the sulci) and separating gray and white matter, we might even craft at least a coarse map of the distribution of neurons along the cortical surface and brain of the African elephant. We had previously used a deli slicer to turn a human brain hemisphere into one such full series of thin cuts. The slicer was wonderful for separating cortical gyri—but it had one major drawback: too much of the human brain matter stayed on its circular blade, precluding estimates of the total number of cells in the hemisphere. If we wanted to know the total number of neurons in the elephant brain hemisphere, we had to cut it by hand, and in thicker slices, to minimize eventual losses to the point of making them negligible.

And so the day started at the hardware store, where my daughter and I (school vacation having just started) went looking for L-brackets to serve as solid, flat, regular frames for cutting the elephant hemisphere, plus the longest knife I could hold in one hand. (Here was an opportunity not to be missed for a young teenager, who years later could say, "Hey, Mom, remember the day we sliced up an elephant brain?") A journalist friend, Bernardo Esteves, in the lab that day to document the process for a piece on brain soup he was writing for his magazine,[15] watched as we first sawed off the structural reinforcements of the L-brackets then made the elephant brain fit inside. Sure, there are fancy hundred-thousand-dollar machines that would do the job to perfection, but why spend that much money when a handheld butcher knife would do the job well enough?

I laid the hemisphere flat on the benchtop, medial wall down, and made one central, frontal cut first, top to bottom, splitting the hemisphere into a front and a back half, then placed the newly cut back face of the front half down, framed inside the two L-brackets. A student held the frames in position while I held the hemisphere down with my left hand and sliced firmly but gently through the brain with the right, in back-and-forth movements, much like the vibratome that cuts much thinner sections of small brains in the lab. Several cuts later, also into the back half as well as the cerebellum, and we had a completely sliced elephant brain "loaf" lying flat on our benchtop (figure 6.2): sixteen sections through the cortical hemisphere, eight through the cerebellum, plus the entire brainstem and the gigantic, 20-gram olfactory bulb (ten times the mass of a rat brain) lying separately.

Next, we had to separate the internal structures—striatum, thalamus, hippocampus—from the cortex, then cut the cortex into smaller pieces for processing, then separate each of these pieces into gray and white matter. In all, we had 381 pieces of tissue, most of which were still several times larger than the 5 grams we could process at one time. It was by far the most

Figure 6.2
Right hemisphere of the brain of an African elephant cut into sixteen sections (top two rows; the front of the brain is at the top right), and its right cerebellum cut into eight sections (bottom row; the medial part of the cerebellum is to the right). This is a truly large brain: the ruler at the top of the image measures 15 centimeters (6 inches). The gray and white matter of the cerebral cortex and of the cerebellum are clearly distinguishable in the sections.

tissue we had processed. One person working alone and processing one piece of tissue per day would need well over one year—nonstop—to finish the job. This clearly had to be a team effort, especially if I wanted to have the results in no more than six months. But, even with a small army of undergraduates, led by senior undergrad Kamilla Avelino-de-Souza, it was taking too long: two months went by and only one-tenth of the brain hemisphere had been processed. Something had to be done.

Capitalism came to the rescue. I did some math (one of the few advantages of doing the accounting for my own grants) and realized I had some 2,500 U.S. dollars to spare—roughly one dollar per gram of tissue to be processed. I gathered the team and made them an offer: anybody could help, as long as either Kamilla or I was in the lab to supervise, and everyone would be rewarded financially by the same amount. I would start keeping a log of who processed what tissue, and quality checks would of course be in place—but the students doing most of the counts, Kamilla and Kleber Neves, were experienced counters, and Kamilla in particular had no qualms about nagging the younger students to do the job right and making them take a poorly dissociated suspension back to the grinder again and again until it was done properly. Small partnerships quickly formed, with one student doing the grinding, the other doing the counting, and both sharing the proceeds. It worked wonders. My husband would visit the lab and comment, in awe, on the crowd of students at the bench, chatting animatedly while working away (until then, they mostly worked in shifts, it being a small lab). Jairo Porfírio took over the large batches of antibody stains, I did all the neuron counts at the microscope—and in just under six months we had the entire African elephant brain hemisphere processed, as planned.

And the Winner Is ...

Lo and behold, the African elephant brain *had* more neurons than the human brain.[16] And not just a few more: a full *three times* the number of neurons, 257 billion to our 86 billion neurons. But—and this was a huge, immense "but"—a whopping 98 percent of those neurons were located in the cerebellum, at the back of the brain. In every other mammal we had examined so far, the cerebellum concentrated most of the brain neurons, but never much more than 80 percent of them. The exceptional distribution of neurons within the elephant brain left a relatively meager 5.6 billion neurons in the whole cerebral cortex itself (in both hemispheres: all numbers obtained for the right hemisphere were doubled to yield whole-brain numbers comparable to other species). Despite the size of the African elephant cerebral cortex (an impressive 2.8 kilograms, or over 6 pounds), the 5.6 billion neurons in it paled in comparison to the average 16 billion neurons concentrated in the much smaller human cerebral cortex (1.2 kilograms or 2.6 pounds), and even in comparison to the 9 billion neurons we estimated in the cerebral cortex of the gorilla and the orangutan.[17]

Interestingly, the cerebral cortex of the elephant had just the mass expected for its number of neurons, compared both to other afrotherians like itself and to rodents, eulipotyphlans, and artiodactyls. As seen in figure 6.3, the elephant cerebral cortex fits the same neuronal scaling rules that apply to other nonprimate species. Nothing special there, then: like the human cerebral cortex, a scaled-up primate cortex, the elephant cortex is just a cortex scaled up in the manner characteristic of the group of mammals to which it belongs.

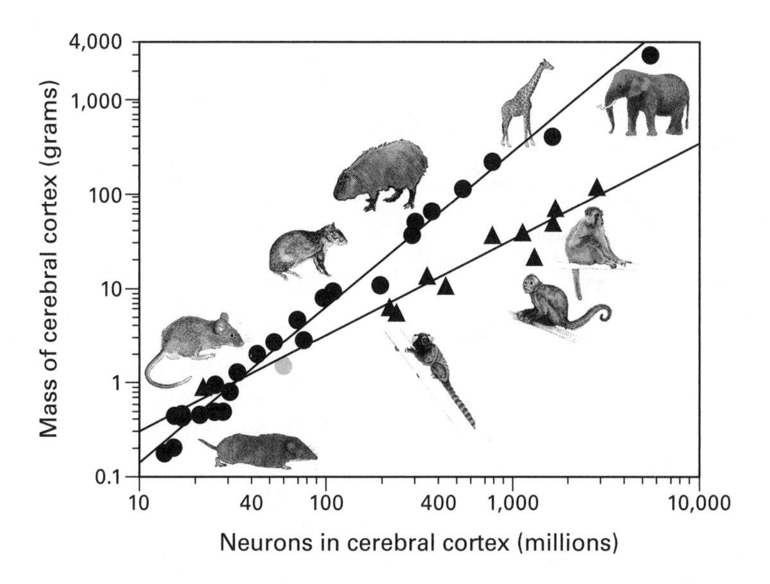

Figure 6.3
The cerebral cortex of the African elephant fits the neuronal scaling rules that apply to nonprimates: its 5.6 billion neurons are very close to the number of neurons predicted for a cortex of its mass.

The African elephant cerebellum, on the other hand, was special in not one, but two ways. First, it had over ten times as many neurons as a generic mammalian cerebellum could be expected to have given the number of neurons in the accompanying cerebral cortex. The average ratio for mammals is 4.2 neurons in the cerebellum to every neuron in the cortex, and that ratio in the elephant was 44.8. Second, that excessive number of neurons in the elephant cerebellum did not fit the neuronal scaling rules for afrotherian and nonprimate, noneulipotyphlan mammals (figure 6.4). For its 251 billion neurons, the elephant cerebellum should weigh over 7 kilograms if it were built like an afrotherian or generic nonprimate cerebellum—that is, be 6.1 times larger in mass than it actually is. The fact that the elephant cerebellum is an outlier in both relationships means that it has an extraordinarily large number of neurons that are extraordinarily

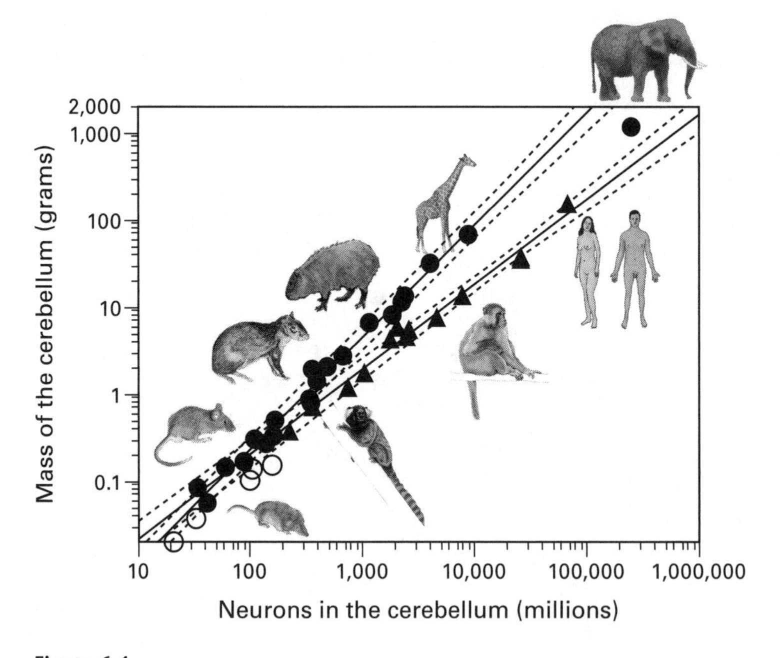

Figure 6.4
The cerebellum of the African elephant has many more neurons than expected for its mass according to the scaling rules that apply to other afrotherians as well as to rodents and artiodactyls (filled circles), nearing (but not quite reaching) the scaling rules that apply to the primate cerebellum.

small in size, suggesting that the elephant cerebellum has undergone positive selection for both an exceptionally large number of neurons and a relatively smaller average neuronal cell size. Not that average neuronal cell size in the elephant cerebellum is exceptionally small, though; given the unexceptional neuronal density of 214,000 neurons per milligram found in the elephant cerebellum, the estimated average neuronal cell size in the structure is comparable to that found in the pig. What is exceptional is that the average neuronal density is much higher than expected given the elephant cerebellum's very large number of neurons, which means that its neurons failed to increase in size accordingly. This explains why the elephant cerebellum, at 1.2 kilograms, is relatively large (it accounts for 25 percent of the elephant's total brain mass, in contrast to the typical 10–15 percent in other mammals), but still not as large as the over 7 kilograms that it should weigh if it scaled according to the rules that apply to other afrotherians.

Why so many neurons in the elephant cerebellum? The cause must be something that is both unique to elephants and requires great numbers of cerebellar neurons. There are two likely candidates at this point: infrasound communication and dealing with information from the trunk, both of which require the processing of information that travels up and down the trigeminal nerve, which feeds into the cerebellum among other brain structures, and which is indeed enormous in the African elephant, almost as thick as its spinal cord. Specialized communication has also been proposed as the cause for the relatively large cerebellum found in echolocating microbats and cetaceans.[18] However, we have recently found that the cerebellum of microbats does not have a disproportionate number of neurons like that found in the elephant cerebellum; rather, their cerebellum is relatively large in mass within the brain only because the microbat cerebral cortex, for its number of neurons, is relatively small.[19] This finding, although in an unrelated group of mammals, suggests that specialized communication by itself does not require a relatively enlarged number of neurons in the cerebellum. That leaves the highly sensory elephant trunk, a 100-kilogram muscular organ capable of precise, delicate movements, as the most likely source of selective pressure for a much larger number of neurons in the elephant cerebellum than might otherwise be expected.

Such a directly sensorimotor function of the cerebellum, without involving the cerebral cortex, would also explain the break away from the tight linear relationship between numbers of neurons in the cerebral cortex and cerebellum in other species. The elephant's larger number of cerebellar neurons might indeed be expected in order to deal with both the parallel processing of information from the cerebral cortex and from the trigeminal nerve related to the trunk.

So here was our answer. No, the human brain does not have more neurons than the much larger elephant brain—but the human *cerebral cortex* has nearly three times as many neurons as the over twice as large cerebral cortex of the elephant. Unless we were ready to concede that the elephant, with three times more neurons in its cerebellum (and, therefore, in its brain), must be more cognitively capable than we humans, we could rule out the hypothesis that total number of neurons in the *cerebellum* was in any way limiting or sufficient to determine the cognitive capabilities of a brain.

Only the cerebral cortex remained, then. Nature had done the experiment that we needed, dissociating numbers of neurons in the cerebral cortex from the number of neurons in the cerebellum. The superior cognitive capabilities of the human brain over the elephant brain can simply—and only—be attributed to the remarkably large number of neurons in its cerebral cortex.

While we don't have the measurements of cognitive capabilities required to compare all mammalian species, or at least those for which we have numbers of cortical neurons, we can already make a testable prediction based on those numbers. If the absolute number of neurons in the cerebral cortex is the main limitation to the cognitive capabilities of a species, then my predicted ranking of species by cognitive abilities based on number of neurons in the cerebral cortex would look like this:

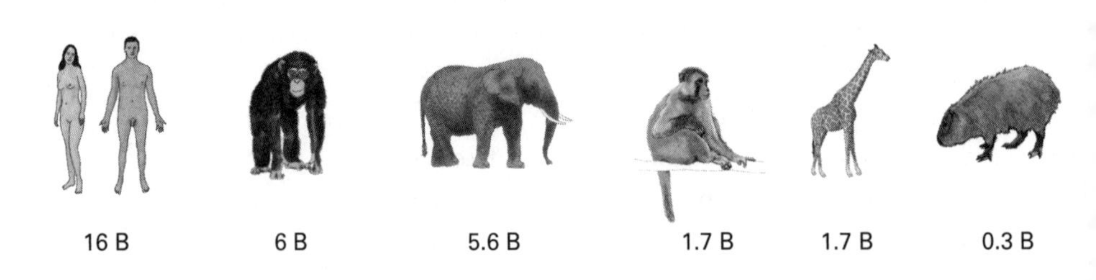

which is more intuitively reasonable than the current ranking based on brain mass, which places artiodactyls (notably, the giraffe) above many primate species, like this:

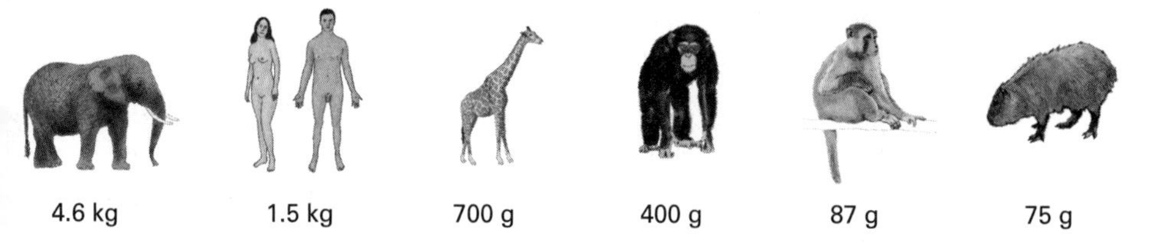

What about Cetaceans?

Cetaceans have been celebrated as animals of such cognitive prowess that some even advocate that they be granted the status of nonhuman persons and, as such, no longer be kept in captivity.[20] Due to the public appeal of these animals, caution in interpreting their behavior is often thrown to the wind. I have seen documentaries in which would-be "dolphin communicators" marvel at how dolphins are interested in humans, when what I see on the screen is a dolphin being chased by said humans. My colleague and collaborator Paul Manger[21] is one of the most notoriously vocal researchers to remind us of the danger of overinterpreting the actions of dolphins in particular. For instance, although the act of nudging drowning humans to the surface has been hailed as evidence of empathy and even a feeling of kinship toward our species, this is a behavior dolphins show not only to their newborn calves (who, like their parents, must surface to breathe, mammals that they are) but also to inanimate objects sinking in the water, and therefore might be extended to drowning members of our species simply out of their similarity to sinking calves or objects. As both Manger and psychologist Onur Güntürkün point out, dolphins, like humans, are not alone in their cognitive capabilities, some of which are even shared with birds and dogs, such as understanding language and referential pointing. If a particular range of complex behaviors is to be used to grant dolphins a higher cognitive status, then other species should be granted the same privileges on the same basis.

Still, several reports on the bottlenose dolphin (*Tursiops truncatus*), the most studied of all cetaceans, have shown that even though, unlike primates and even birds, dolphins do not seem to make use of objects as tools, they do recognize themselves in the mirror,[22] like humans and other large primates,* and even make decisions based on a distinction between "few" and "many" elements in a pattern, which they learn to categorize through experience.[23] Moreover, dolphins seem to have something akin to "names" for themselves; each dolphin has a signature whistle and upon hearing it produced by another dolphin or loudspeaker, it calls back.[24] And, just as humans can remember other humans' names even after decades apart, so dolphins can recognize the signature whistles of their conspecifics over time spans of at least twenty years.[25]

There are a few cetacean species that have even larger brains than the African elephant. Examining the neuronal composition of their brains will be the final test to the hypothesis that the 16 billion neurons in the human cerebral cortex, more than in any other cortex regardless of its size, are the simplest basis for our outstanding cognitive abilities. That is something that we are working on right now. And although the numbers aren't in yet, there are some informed predictions that can be made.

Cetaceans are actually close cousins of artiodactyls, grouped together in the order Cetartiodactyla, in which case the neuronal scaling rules that we found to apply to the cerebral cortex of artiodactyls (and rodents, eulipotyphlans, and afrotherians) should also extend to whales and dolphins. We can thus predict that the large cerebral cortex of several cetacean species, such as the pilot whale (*Globicephala macrorhyncha*), which is about twice as large as the human cerebral cortex, would still be composed of only around 3 billion neurons. Even the largest whale cortex, at 6–7 kilograms, is predicted to still have fewer than 10 billion neurons (figure 6.5).

The not-so-large number of neurons that we predict for the cortex of large whales, based on the scaling rules for their cousin artiodactyls, is at odds with the much larger previous stereological estimate of 13 billion neurons in the cerebral cortex of the Minke whale[26] (*Balaenoptera acutorostrata*), and the even larger estimate of 15 billion neurons in the smaller cortex of the harbor porpoise[27] (both still smaller than the number of neurons in the human cortex, though barely). However, both these studies suffered from the same unfortunately common problem in stereology: undersampling, in

*The self-recognition ability of dolphins has been questioned, however, by Paul Manger (2013) and Onur Güntürkün (2014).

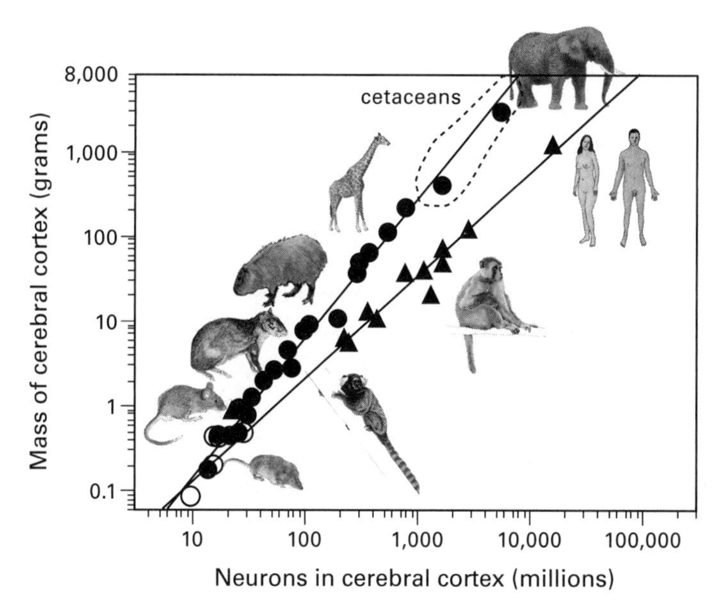

Figure 6.5
The cetacean cerebral cortex (values within dashed oval) is expected to share the neuronal scaling rules that apply to the closely related artiodactyls and to the more distant rodents, afrotherians, and eulipotyphlans (circles), and thus to hold a smaller number of neurons than a smaller primate cortex such as the human.

one case drawing estimates from only 12 sections out of over 3,000 sections of the Minke whale's cerebral cortex, sampling a total of only around 200 cells from the entire cortex, when it is recommended that around 700–1,000 cells be counted per individual brain structure.[28] With such extreme undersampling, it is easy to make invalid extrapolations—like trying to predict the outcome of a national election by consulting just a small handful of people.

It is thus very likely, given the undersampling of these studies and the neuronal scaling rules that apply to cetartiodactyls, that even the cerebral cortex of the largest whales is composed of only a fraction of the average 16 billion neurons we found in the human cerebral cortex. The human advantage, I would say, lies in having the largest number of neurons in the cerebral cortex that any animal species has ever managed—and it starts by having a cortex that is built in the image of other primate cortices: remarkable in its number of neurons, but not an exception to the rules that govern how it is put together. Because it is a primate brain—and not because it is special—the human brain manages to gather a number of neurons in a still comparatively small cerebral cortex that no other mammal with a viable brain, that is, still smaller than 10 kilograms, would be able to muster. Figure 6.6, in a semilog scale, illustrates that distance between humans and the next tier in numbers of neurons in the cerebral cortex: great apes, the African elephant, and, we predict, large cetaceans.

It seems fitting that great apes, elephants, and probably cetaceans have similar numbers of neurons in the cerebral cortex, in the range of 3 to 9 billion: fewer than humans have, but more than all other mammals do. These are, many scientists think, the three groups of mammalian species that share a number of complex cognitive capabilities such as self-recognition in the mirror, vocal communication, and social cooperation to achieve a common purpose.

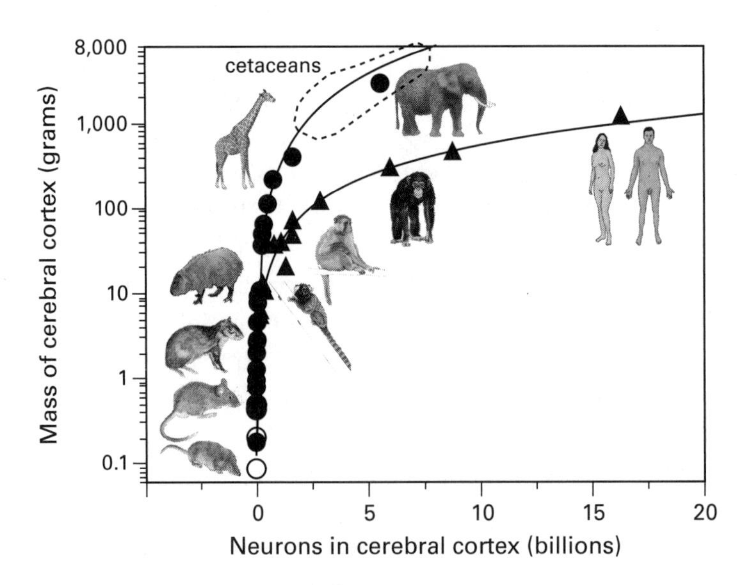

Figure 6.6
Same data as in figure 6.5 plotted on a semilog scale to make it easier to appreciate the distance between the number of neurons in the cerebral cortex of humans and in that of other mammals. Cetaceans (values within dashed oval), the African elephant, and great apes are estimated to share the same range of numbers of neurons in the cerebral cortex—between 3 and 9 billion neurons—which is well beyond the number contained in the cortex of other, smaller mammalian species, but still far fewer than found in the cerebral cortex of humans.

One could still argue that the human cerebral cortex is special, yes, in that it concentrates a number of neurons so large that it is unreachable for other species. How we managed to have this unprecedented and largest number of neurons in the cerebral cortex without ever breaking away from the primate way of building a cortex is the subject of the final chapters of this book. In the meantime, there is a particular point to consider as part of the human advantage: not only do we have the largest absolute number of neurons in the cerebral cortex as a whole, but we might have a particularly enlarged cortex relative to the rest of the brain and a particularly enlarged prefrontal cortex within it.

But do we really?

7 What Cortical Expansion?

When we began to study the scaling of the number of neurons in the brain, mammalian evolution had long been synonymous with expansion of the cerebral cortex because of the prior emphasis on brain volumes, the only data available for decades. Human evolution, in particular, was supposed to have been largely an affair of cortical expansion: even though the human brain is not the largest in absolute mass or volume, it has the largest cerebral cortex relative to total brain mass. Not by much, however. Including the subcortical white matter, with the fibers that connect the different cortical locations and allow all of them to work as one, the cerebral cortex had been reported to represent around 75 percent of brain mass or volume in the human, against 74 percent in the horse, 71–73 percent in the chimpanzee, and 73 percent in the short-finned pilot whale.[1]

Our own data set confirmed the tendency for the mammalian cerebral cortex to gain mass faster than all other brain structures with increasing brain size—and, most important, faster than the rest of brain, as shown in figure 7.1, where the plotted functions have allometric exponents greater than 1. This meant that, as the structures in the rest of brain that process information from the body became larger, the cerebral cortex grew even more. As should be no surprise by now, we found that primates differed from the other mammalian groups also in this respect: the cerebral cortex became larger relative to the rest of brain faster than it did in other species. Additionally, for a similar mass of the rest of brain, the cerebral cortex was larger in primates than in nonprimate species, as seen from the separation between the two plotted lines in figure 7.1. More important, and confirming our conclusion that the human brain was simply a scaled-up primate brain, the mass of the human cerebral cortex was just what would be

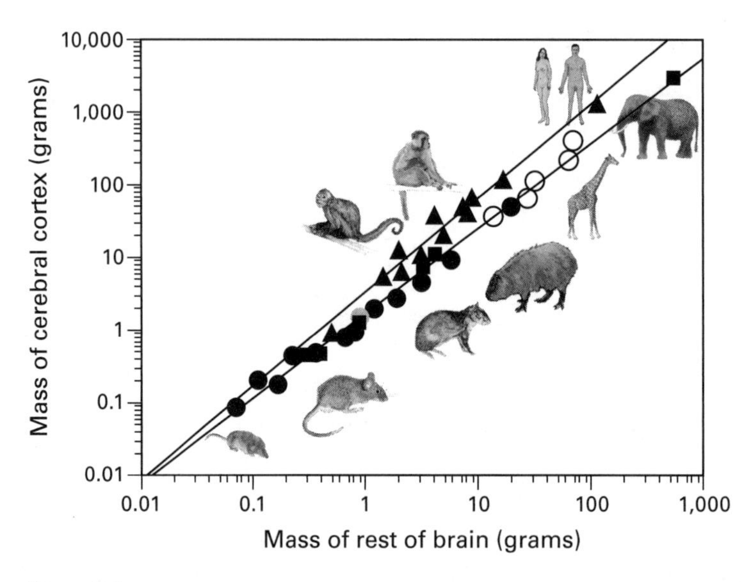

Figure 7.1
The mass of the cerebral cortex increases faster than the rest of brain in all mammals, but particularly faster in primates (exponent, 1.3; triangles) than in other species (exponent, 1.2). Additionally, for a similar mass in the rest of brain, primates (triangles) have a larger cerebral cortex than nonprimates (rodents and eulipotyphlans, filled circles; artiodactyls, open circles; afrotherians, squares). The human cerebral cortex has the mass predicted to accompany the mass of its rest of brain for a primate.

predicted for a generic primate brain having the same mass in its rest of brain.

Considering that the rules that apply to nonprimates as a whole have been conserved in evolution, and these rules therefore also applied to the ancestral mammal common to all these groups, we can infer that the cerebral cortex has been gaining mass faster than the rest of brain ever since the first mammals appeared—and that primates diverged away from that pattern with a stepped-up increase in the mass of the cerebral cortex, and an even faster increase in the mass of the cerebral cortex relative to the rest of brain. This means that both the absolute expansion of the cerebral cortex and its relative increase over the rest of brain have been particularly fast in primate evolution.

With the exponents of its power functions significantly larger than unity, cortical expansion in the evolution of mammals, then, has been both absolute and relative, compared to the rest of brain. As shown in figure 7.2, the relative mass of the cerebral cortex, that is, the percentage of brain mass contained in the cerebral cortex, increases together with brain mass across species within each group of mammals—but differently across primates and nonprimates, such that the antelope has 72 percent of its brain mass in the cerebral cortex, compared to 80 percent for the baboon. The African elephant, with its oversized cerebellum, stands out: although its brain is about three times the size of the human brain, the elephant cerebral cortex represents only 62 percent of brain mass, compared to the human cerebral cortex, which represents 82 percent.

Because the first mammals to evolve were small, the increasing relative mass of the cerebral cortex with increasing brain mass is the basis for the notion that, in the course of its evolution, the cortex has undergone both

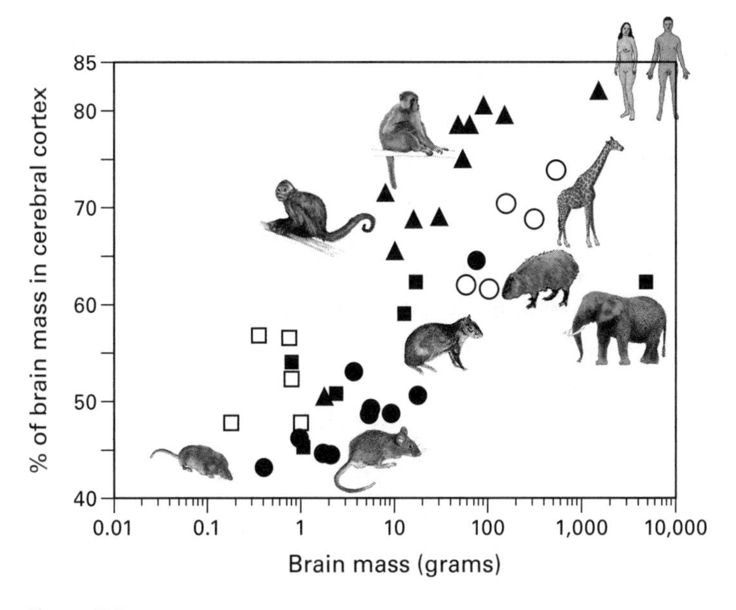

Figure 7.2
Larger brains have relatively larger cortices, but the relative mass of the cerebral cortex is different across mammalian groups. The cortex is relatively larger in eulipotyphlans (open squares) than in afrotherians (filled squares) and rodents (filled circles) of similar brain mass, and larger in primates (triangles) than in artiodactyls (open circles) and rodents of similar brain mass. The relative mass of the cerebral cortex is largest in humans than in all other species in our dataset.

an absolute and a relative expansion, slowly taking over brain mass—and, supposedly, brain functions as well[2]—in primates and nonprimates alike. This is an important concept: because the cerebral cortex receives information from the rest of brain and elaborates on it, adding a whole other level of complexity, an expansion of the cerebral cortex relative to all other parts of the brain, in principle, means that the larger brains become, the more capable they are of complex and flexible functions and behaviors beyond simply operating the body.

And at the pinnacle, supposedly, is the human cerebral cortex, with the largest relative size compared to the brain. That, however, is only to be expected, both because we are primates and because, among primates, we have the largest brain and cerebral cortex, not because we are special. So, again, humans are just the continuation of an evolutionary trend.

In any case, this "taking over" of brain functions by the ever larger cerebral cortex presupposes that a relatively larger cerebral cortex has a relatively larger number of neurons. This had been the basic hypothesis behind the importance attributed to cortical expansion in evolution—but, for lack of data on numbers of neurons, it had yet to be tested.

Now that we had the data, we could test whether larger brains with disproportionately larger cortices did indeed have disproportionately larger numbers of neurons in the cerebral cortex, consistent with the evolutionary takeover of brain functions by the cerebral cortex—and, in particular, whether our human cerebral cortex had not only the largest absolute number of neurons, but also the largest number of neurons relative to the whole brain.

And here we were in for a big surprise: the answer, in both cases, was No. If relatively larger cortices also had relatively more neurons, there should be a positive correlation between the percentage of brain mass and the percentage of brain neurons in the cerebral cortex. But there is none: as shown in figure 7.3, there is a fair amount of variation, but in most mammalian brains, regardless of its relative size, the cerebral cortex holds the same 15 to 25 percent of all brain neurons. The human cerebral cortex, in particular, holds only 19 percent of all brain neurons, even though it amounts to 82 percent of all brain mass in our data set. This is the same as the cerebral cortices of both the guinea pig and capybara, which also hold 19 percent of brain neurons, although they amount to only 53 percent and 64 percent of brain mass, respectively, compared to our 82 percent. Thus cortical

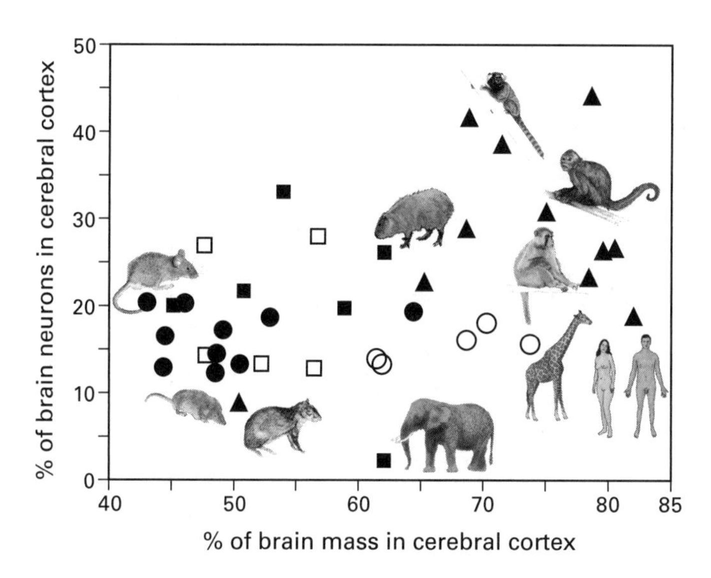

Figure 7.3
Relatively larger cortices do not concentrate relatively more of all brain neurons. There is a fair amount of variation, but in most species, the cerebral cortex holds 15–25 percent of all brain neurons, regardless of its relative mass, in eulipotyphlans (open squares), afrotherians (filled squares), rodents (filled circles), artiodactyls (open circles) and primates (triangles) alike. The main exception is the African elephant, whose cerebral cortex holds only 2 percent of all brain neurons.

expansion in mammalian evolution has been about relative and overall mass, but *not* about relative number of neurons. Even though a larger cerebral cortex does have more neurons within each mammalian group, a *relatively* larger cerebral cortex does not have relatively more neurons within the brain as a whole.

How could this be? How could the cerebral cortex become relatively larger but *not* have relatively more neurons as it expanded within the brain? Mathematically, some other part of the brain had to be gaining neurons together with the cerebral cortex, but gaining mass more slowly than the cortex did. This structure turned out to be the cerebellum. As shown in figure 7.4, the mammalian cerebellum accounts for around 80 percent of all brain neurons—except in the African elephant, whose cerebellum concentrates 98 percent of all brain neurons.

When plotting variations in the number of neurons in the cerebellum against the number of neurons in the cerebral cortex, what we found was

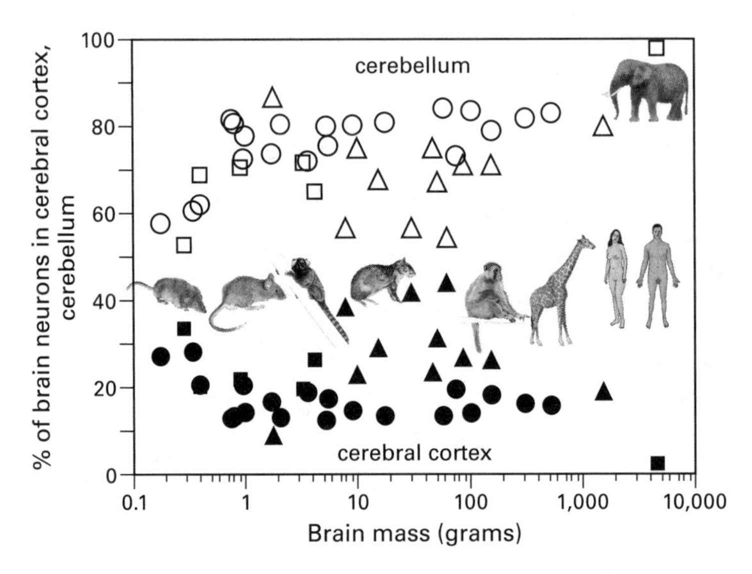

Figure 7.4
The cerebellum (open symbols) holds around 80 percent of all brain neurons in most mammalian species, while the cerebral cortex (filled symbols) holds 15–20 percent of all brain neurons in eulipotyphlans, rodents, and artiodactyls (circles); afrotherians (squares; except for the elephant); and primates (triangles). The main outlier is the elephant, whose cerebral cortex holds only 2.2 percent of all brain neurons, with 97.5 percent of all brain neurons in the cerebellum and only 0.3 percent in the rest of brain.

not that the numeric preponderance of cortical over cerebellar neurons increased with larger brains across species, but, instead, that the two brain structures gained neurons proportionately in a linear relationship, such that, on average, the cerebellum gained four neurons for every neuron added to the cerebral cortex (that is, the function plotting the linear relationship between cerebellum and cerebral cortex in figure 7.5 has a slope of around 4.0).

Because of the different neuronal scaling rules that apply to the cerebral cortex and the cerebellum (remember that the average mass of neurons in the cerebral cortex scales faster than that in the cerebellum in nonprimate species), the cortex gains in mass (and volume) much faster than the cerebellum, thus becoming relatively larger than the cerebellum (and the rest of brain) even though the ratio of cerebellar to cortical neurons remains constant, with the cerebellum gaining four neurons proportionately for every

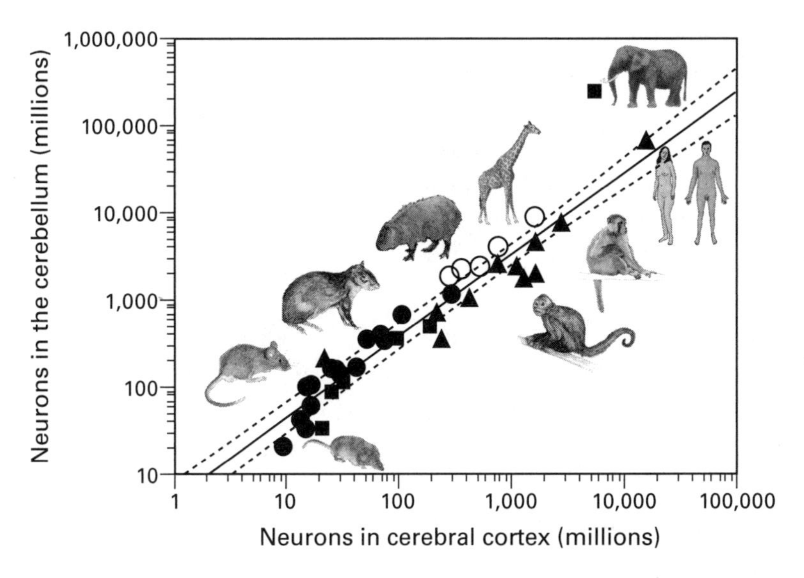

Figure 7.5
Cerebellum and cerebral cortex gain neurons proportionately, with, on average, four neurons added to the cerebellum for every neuron added to the cerebral cortex across mammalian species. The obvious exception is the African elephant, which has more than ten times the number of neurons in its cerebellum than predicted for the number of neurons in its cerebral cortex.

one neuron the cerebral cortex gains. Even in primates, the very small difference in the exponents of the neuronal scaling rules that apply to the cerebral cortex and cerebellum is already enough to lead, across several orders of magnitude of variation in the numbers of neurons in these structures, to the increase in relative cortical mass in larger brains. This coordinated, linear, proportionate addition of neurons to the cerebral cortex and cerebellum of different species, combined with the different scaling rules for each structure, explains how the cerebral cortex becomes increasingly larger in mass relative to the brain and yet maintains a fairly constant 15–25 percent of all brain neurons.

The coordinated, linear scaling of numbers of neurons in the cerebral cortex and cerebellum of most mammalian species has a fundamental functional implication: that mammalian evolution has *not* been about a takeover of brain functions by the cerebral cortex alone. Rather, in all species but the elephant, the cerebral cortex and the cerebellum have worked in tandem, such that any increment in processing capabilities of the cerebral cortex has been accompanied by a proportionate increment in the processing capabilities of the cerebellum.* This is all the more obvious in the human brain: its cerebral cortex concentrates the largest absolute number of neurons in any mammalian brain—but still has the same relative percentage of brain neurons as other mammals, elephant excluded.

Cortical Expansion over the Rest of Brain

We had found, then, that the expansion of relative cortical mass in mammalian evolution did not reflect an expansion in the relative number of neurons in the cerebral cortex; rather, the cerebral cortex and cerebellum gained neurons proportionately, such that the percentage of brain neurons in each—between 15 and 25 percent and 75 and 80 percent, respectively—remained fairly constant across species. But it was still possible that the cerebral cortex and the cerebellum did indeed gain neurons faster than the rest of brain, thus supposedly adding complexity to the processing of sensorimotor information to and from the body. Knowing the numbers of neurons that formed the two structures, it was an easy hypothesis to test.

Across nonprimate, nonartiodactyl species, we found that, much to the contrary of what the relative increase in cortical mass suggested, the cerebral cortex gained neurons linearly with the rest of brain, as shown in figure 7.6. These species maintained a constant ratio of two neurons in the cerebral cortex (and eight in the cerebellum) for every neuron in the rest of brain (figure 7.7), which we have proposed to be the ancestral way of distributing neurons across the main brain structures.[3]

Primates, in contrast, did show a much faster addition of numbers of neurons to the cerebral cortex compared to the rest of brain: here, the exponent of the power function in figure 7.6 is 1.4, significantly larger than unity. This means that, as primate brains added neurons to the rest of brain, they added even more neurons to the cerebral cortex, such that the ratio of neurons in the cerebral cortex to neurons in rest of brain increases

*The coordinated addition of numbers of neurons to the cerebral cortex and cerebellum corroborates evidence from functional MRI studies that the cerebellum participates in all aspects of cognitive function that involve the cerebral cortex (see Ramnani, 2006).

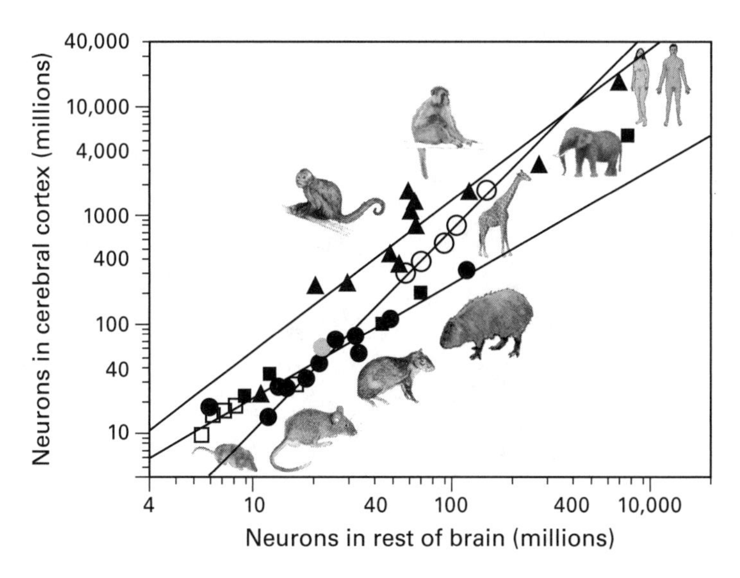

Figure 7.6
Number of neurons in the cerebral cortex scales linearly with the number of neurons in the rest of brain across eulipotyphlans (open squares), afrotherians (filled squares, excluding the African elephant), and rodents (filled circles), but it scales with the number of neurons in the rest of brain raised to the power of 1.9 in artiodactyls (open circles) and to the power of 1.4 in primates (triangles). Thus, both primates and artiodactyls gain cortical power of information processing faster than they gain neurons in the rest of brain to convey information to be processed.

(figure 7.7), reaching up to 27 to 1 (or, more simply, a ratio of 27), supposedly increasing the complexity and flexibility of information processing by the cortex.

Surprisingly, however, primates were not alone: artiodactyls also showed an increased proportionate allocation of neurons to the cerebral cortex over the rest of brain, with an even steeper exponent of 1.9 (figure 7.6). This translates into ratios of neurons in the cerebral cortex to neurons in the rest of brain of between 5 (in the pig) and 11 (in the giraffe), ratios that are intermediate between those for primates and those for other nonprimate species (figure 7.7). The elephant also seems to be an outlier, with about three times the number of neurons in its cerebral cortex expected for an afrotherian with its number of neurons in the rest of brain.

Thus, as both primates and artiodactyls gained larger brains, their cerebral cortices gained neurons disproportionately over the rest of brain,

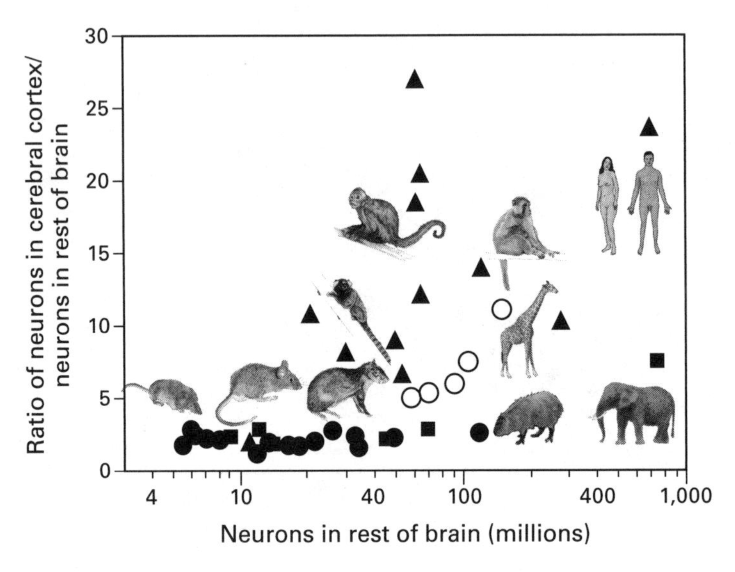

Figure 7.7
There are on average two neurons in the cerebral cortex for every neuron in the rest of brain in eulipotyphlans (filled circles), afrotherians (squares, excluding the African elephant), and rodents (filled circles), but between 5 and 11 cortical neurons for every neuron in the rest of brain of artiodactyls (open circles) and between 2 and 27 in primates (triangles). The human brain has 24 neurons in its cerebral cortex for every neuron in its rest of brain, although the maximal ratio is found in the bonnet monkey brain.

supposedly increasing the degree of complex processing of information relayed to and from the body. This would have conferred to primates and artiodactyls quite an advantage over other mammals, such as rodents and afrotherians, in flexible and complex processing of body-related information. But because, at the same time, the cerebellum in primates and artiodactyls gained neurons proportionately with the cerebral cortex, and thus disproportionately relative to the rest of brain, the disproportionate increase in cortical neurons over neurons in the rest of brain occurred *without* a corresponding disproportionate increase in cortical neurons over neurons in the brain *as a whole*.

Interestingly, the ratio of neurons in the cerebral cortex to neurons in the rest of brain is high but not maximal in humans: there are 24 neurons in the human cerebral cortex for every neuron in its rest of brain, but 27 in the bonnet monkey. Thus, unless we are willing to accept that this slightly higher ratio has conferred slightly greater brain processing power to the bonnet monkey cortex and thus also a slightly greater cognitive advantage to bonnet monkeys over humans, the most likely explanation for the cognitive advantage of humans over other species remains the remarkably large absolute number of neurons in our cerebral cortex—ten times the number in the bonnet monkey cortex.

An Expanded Prefrontal Cortex?

One of the pervasive ideas in human brain evolution has been not simply that our cerebral cortex as a whole is relatively larger than in other species, but that, within this enlarged cerebral cortex, our prefrontal areas—those areas located at the front of our cortex, required for planning and for logical reasoning[4]—are also relatively larger. This notion dates back to the German neurologist Korbinian Brodmann, who in the early twentieth century mapped cortical areas in human, chimpanzee, and macaque brains. Brodmann estimated that the frontal cortex, the part of the brain that contains the associative, prefrontal areas,* occupies 29 percent of the cerebral cortex in the human, but 17 percent in the chimpanzee, and only 11 percent in the rhesus monkey.[5]

It took nearly one century and the advent of modern neuroanatomical techniques that allowed the brain to be measured while still inside living animals for those numbers to be challenged. But Katerina Semendeferi and colleagues[6] showed in 2002 that the frontal cortex of humans, bonobos, chimpanzees, gorillas, and orangutans occupies the same 35 to 37 percent of all cortical volume. Although larger than the 30 percent of cortical volume occupied by frontal areas in macaques and other smaller primates, it in no way singles out the human prefrontal cortex. The cognitive superiority of human brains could no longer be attributed to a relatively enlarged

*There is a small but important distinction: the prefrontal cortex is the associative (that is, non-motor) part of the frontal cortex. Distinguishing the *frontal* cortex is straightforward in primates: it is all the cortex that is anterior to the central sulcus, the main trough in the brain surface. But distinguishing the prefrontal cortex within it is a much more complicated matter that requires fine anatomical and functional analyses. So, while the non-motor, prefrontal cortex is the part strictly associated with higher cognition, it is much more feasible to measure and compare the frontal cortex across primate species.

frontal cortex. Maybe a richer connectivity was the cause, the authors pondered.

Indeed, one of the studies that followed, by Thomas Schoenemann and colleagues,[7] found that while the human prefrontal cortical gray matter* had just the relative volume expected for a primate cortex of its total mass, the prefrontal white matter, which holds all those fibers connecting prefrontal areas to themselves and to other locations in the cortex, appeared to be relatively larger in humans than expected for the relative volume of the corresponding gray matter. Another study, by Jeroen Smaers and colleagues,[8] corroborated a relative expansion of the human prefrontal white matter—and only in the left hemisphere, which houses speech production. There it was again: the white matter of the human prefrontal cortex was out of the ordinary.

The problem was that, when Robert Barton and Chris Venditti[9] independently subjected the data of Schoenemann and of Smaers to a slightly different form of mathematical analysis, taking phylogenetic proximity into consideration and also taking care not to confound individual variation with variation across species, there was no significant difference in the relative volume of the human prefrontal white matter compared to that expected for a primate of our brain mass. Regardless of the controversy that raged as specialists took sides in the debate, it appeared that any possible exceptionality of the white matter volume of the human prefrontal cortex must be small since it depended on the statistical analysis applied.

As with all other previous studies, there was the additional problem that Semendeferi, Schoenemann, Smaers, and Barton had been limited to the use of the one parameter available then: cortical volume—which we already knew was not a simple, direct reflection of numbers of neurons across cortical areas. Within an individual cortex, the density of neurons can vary by a factor of 5 depending on the cortical location.[10] To determine whether the human prefrontal cortex was enlarged compared to other cortical areas, what we really needed to know was whether it had more neurons than expected for the number of neurons in other cortical areas—and whether the underlying white matter was any larger than expected for the number of prefrontal cortical neurons it connected.

*"Prefrontal" was defined here as all cortex anterior to the corpus callosum, an important, practical shorthand that we were to also use later.

This became the Ph.D. thesis work of Mariana Gabi in the lab, once more in collaboration with Jon Kaas. Because there were no reliable anatomical criteria to define the prefrontal cortex across the various human and nonhuman primate species that we wanted to compare, we resorted to the same simple criterion used by Schoenemann: defining prefrontal cortex as all gray and white matter regions anterior to the corpus callosum, the thick fiber bundle that connects the two hemispheres. Though not a perfect criterion, it was appropriate enough, for it was the one that Schoenemann and colleagues had used to determine that human prefrontal white matter was enlarged compared to that of other primates.

What we found when numbers of prefrontal neurons were considered, however, was that, once again, the human cerebral cortex was no different from what was expected. In comparison to seven other primate species, the human prefrontal region had (1) the number of neurons expected for its gray matter volume and for the total number of neurons in the remainder of the cerebral cortex; (2) the white matter volume expected for the number of prefrontal neurons; and (3) the white matter volume and number of neurons expected for the volume and number of neurons in the nonprefrontal subcortical white matter. Our findings supported those of Barton and Venditti: the human prefrontal cortex is no larger than it "should" be.[11]

Nor does it have more neurons, proportionately, than other primate prefrontal cortices: we found the same 8 percent of all cortical neurons located in prefrontal regions, anterior to the corpus callosum, in humans and in other primate species. Actually, we had found earlier that the mouse also has 8 percent of all its cortical neurons situated in associative, prefrontal-like structures.[12] The similarity suggests the possibility that there is a uniform distribution of cortical neurons into sensory, motor, and associative functions across mammalian species—something that we are now looking into.

So the question remained: if the prefrontal cortex has the same relative size and the same proportion of all cortical neurons across human and nonhuman primate species alike, what accounts for our cognitive advantage? "Connectivity," had already suggested Katerina Semendeferi, Robert Barton, and Chris Venditti.

There are essentially two ways in which a "different connectivity" of the human prefrontal cortex might underlie our cognitive advantage. One is that the areas connected to the prefrontal cortex might be different between humans and other species, resulting in more elaborate features emerging from the human prefrontal cortex. Remarkably, though, the layout of which cortical areas are connected to which other cortical areas, that is, the large-scale cortical "connectome," or "wiring diagram" of neural connectivity within the cortex, seems strikingly similar across species as distant as humans, macaques, cats, and even pigeons, with similar hubs concentrating connectivity in the hippocampus and prefrontal-like structures.[13] The human connectome is therefore not fundamentally different from that of other species.

A "different connectivity" of the human prefrontal cortex could also mean more prefrontal synapses compared to other mammalian species. Although actual data to allow a proper comparison are still lacking, what little evidence there is suggests that the density of synapses in the cortical gray matter is fairly constant across mammalian species.[14] If this turns out to be the case, then the larger the cerebral cortex, the more synapses it has—and, since the human cerebral cortex is nowhere near being the largest, the number of our synapses, prefrontal or otherwise, should also be nowhere near the largest. So, back to square one.

But another possibility remained: that, even without a relatively expanded prefrontal cortex, the absolute number of prefrontal neurons in the human cerebral cortex might far exceed that in the cortex of other primate species. The same 8 percent of all cortical neurons amounts indeed to a much larger absolute number of prefrontal neurons in the human cerebral cortex than in that of other primates: 1.3 billion prefrontal neurons in the human cortex, but only 230 million in the baboon, 137 million in the macaque, and a meager 20 million in the marmoset. Given that prefrontal neurons are those which, because of their wide-scale connectivity to other brain areas, are able to perform associative functions, adding complexity and flexibility to behavior and making planning for the future possible, all these capabilities must increase together with the number of neurons available to perform them—just as adding more processors to a computer improves its computational power.

The ability to plan for the future, a signature function of prefrontal regions of the cortex, may be key indeed. According to the best definition I have come across so far, put forward by MIT physicist Alex Wissner-Gross, intelligence is the ability to make decisions that maximize future freedom of action—that is, decisions that keep the most doors open for the future.[15] If this is the case, then intelligence should rely on the ability to use past experience to represent future states (a function of the hippocampus, which gains neurons together with the remainder of the cortex), to plan the appropriate sequence of actions to get there, and to orchestrate their execution (functions of the prefrontal cortex). That is, the larger the number of prefrontal and hippocampal neurons available, the more intelligent a species should be.

It is not yet clear what percentage of all cortical neurons have associative, prefrontal-like functions in large elephant and cetacean brains—but there is evidence that prefrontal areas are but a sliver of their cortical volume.[16] Because of our already larger total number of cortical neurons compared to that of species having even larger cortices, we can safely expect that our 1.3 billion neurons dedicated to associative, prefrontal functions will be matched by none. Once again, our cognitive advantage over other species seems to lie in the sheer number of processing units available to do the job—regardless of the size of the brain mass housing them, as we've seen in previous chapters, or of the body mass controlled by them, as we'll see in the next chapter.

8 A Body Matter?

It comes as no surprise that the largest animals have the largest brains. At the very least, a very large brain could not fit inside a small animal. There is a basic body plan that applies to mammals, which have four limbs and a head always somewhat proportional to the rest of the body. Located inside the head, the brain can only be as large as the head can accommodate. But does the larger head of a larger body always contain a proportionally larger brain? In other words, what is the relationship between brain size and body size?

As we saw in chapter 1, allometry, the science of the proportion of body parts as animals vary from small to large, is at least as old as Galileo Galilei, who recognized that larger animals are not simply uniformly scaled-up versions of smaller animals. Allometry describes the often disproportional scaling of some body parts as the body increases in mass. For example, the infants of any mammalian species look like infants rather than adults because of the different proportionality between head and body size.

Just as there is a basic allometry of body parts such as heads and legs, there is a basic allometry of body organs. In the case of the heart and liver, their scaling is isometric: larger species have proportionally larger hearts and livers, whose mass scales linearly with the body. The logic is that these are organs whose function is related to the volume of the body. The heart pumps a volume of blood that is proportional to the volume of the body, and that task is performed by a mass of cardiac muscle that grows in linear proportion to the volume of blood pumped.[1] Likewise, the liver, which filters the volume of blood pumped by the heart, also grows in linear proportion both to the volume of blood in the body and to the volume of the body itself.

One of the earliest findings in comparative mammalian neuroanatomy was that larger animals tend to have larger brains—but not proportionally larger.[2] Brain mass varies by a factor of nearly 100,000 across mammalian species, from 0.1 gram in the smallest shrews to 9 kilograms in the sperm whale, but body mass varies by a factor of 100 million across mammalian species—more than 1,000 times as much—from a couple of grams in shrews to nearly 100,000 kilograms or even more (200 tons) in the largest whales. As we saw in chapter 1, larger animals have larger brains, yes—but body mass grows faster than brain mass. This means that less and less of the volume of the head is actually occupied by the brain as animals become larger. In the African elephant, for example, the head is so enormous that even a 5-kilogram brain seems to disappear inside it, buried in bone.

The few dozen mammalian species in our data set illustrate this overall tendency for larger bodies to come with only somewhat larger brains (figure 8.1). Across all species in our sample, brain mass scales with body mass raised to the power of +0.774, a scaling exponent significantly smaller than unity, which means that the brain increases in mass more slowly than the body: a 10 times larger animal has a brain that is only some 6 times larger, and a 1,000 times larger animal has a brain that is only 211 times larger. As a consequence, the relative size of the brain compared to the body decreases as animals become larger. It is, nevertheless, a slow decrease: whereas the brain of a mouse accounts for 1 percent of its body mass of some 40 grams, the brain of the 125,000 times larger elephant accounts for 0.1 percent of its body mass of 5,000 kilograms—a 10-fold drop in relative brain mass for a 125,000-fold increase in body mass.

The precise relationship between brain mass and body mass, however, differs across mammalian orders. The scaling exponents in our data set vary from +0.548 in artiodactyls to a nearly linear +0.903 in primates (excluding great apes): again, there is no universal scaling relationship that applies equally well across all mammals, although the general trend is always there. Compared to rodents of a similar body mass, primates always have larger brains. Take, for instance, the guinea pig and the marmoset. Both animals weigh slightly more than 300 grams, but the brain of the guinea pig weighs only 3.6 grams compared to 7.8 grams for the marmoset's brain. In larger animals, the difference is even more striking: the 4.6-kilogram rabbit has a brain of 9.1 grams, but the smaller, 3.3-kilogram capuchin monkey has a brain that weighs 52.2 grams, nearly six times as much. The same applies to eulipotyphlans and afrotherians, animals that were once grouped as

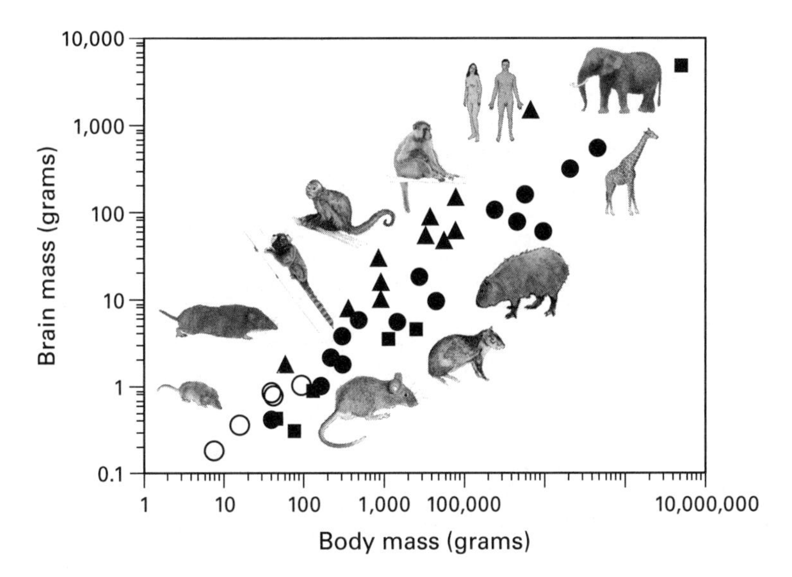

Figure 8.1
Larger body mass is loosely correlated with larger brain mass across mammalian species, but different scaling rules apply to different mammalian orders. For a similar body mass, eulipotyphlans (open circles) have larger brains than afrotherians (squares), and primates (triangles) have larger brains than both rodents and artiodactyls of similar body mass (filled circles). The respective power functions can be found in the appendix.

"insectivores" and still share the words "shrew" and "mole" in many of their common names. Eulipotyphlans, such as the hairy-tailed mole and the European mole, which weigh 42.7 and 95.3 grams, have brains that weigh around 1.0 gram, about twice the size of the brains of afrotherians of similar body mass, such as the elephant shrew and the golden mole, which weigh 45.1 and 79.0 grams.

Still, it remains the case that, within each mammalian group, larger animal species come with larger brains.* One of the rationales, defended by

**Within* a species, though, larger individuals don't necessarily come with larger brains—and the allometric exponent relating brain mass to body mass across individuals of the same species tends to be much smaller than it is across species, if at all significantly different from zero. This fascinating conundrum, which will not, however, be examined here, implies that the scaling of brain and body mass across species in evolution is not simply an extension of the same scaling across individuals of the same species. See Armstrong, 1990, and Herculano-Houzel, Messeder et al., 2015.

Harry Jerison among others, was thought to be that larger bodies needed more neurons to run them, just as more blood required a bigger heart to pump it—although different authors argued that the factor driving the need for more neurons should be either body mass (because of the larger muscular mass to be controlled), body surface (because of the larger sensory surface to be surveyed), or something else altogether. For decades, the debate remained hypothetical because numbers of neurons in the brain were not available to be compared. The best guess until then was that the increase in brain mass was driven by a larger muscular mass, but limited by metabolism: according to studies by Max Kleiber,[3] the energy cost of operating a larger body scales with body mass raised to the power of +0.75, which is fairly close to the scaling exponent relating brain mass to body mass across all mammalian species pooled together (+0.774 in our own dataset), as pointed out by both Harry Jerison[4] and Robert Martin. The logic here was that the increase in brain mass would be limited by how much energy the body could use—again, assuming there was a single universal relationship between brain mass and its energy cost.

We now had the necessary numbers and an answer, which at this point should come as no surprise: there *wasn't* a single, universal scaling rule between body mass and number of brain neurons across all mammals. But I'll get to that in a minute. First, let's start with what we reasoned should be the most direct way to tackle the issue of whether larger bodies required more neurons to operate them: looking at the spinal cord, that part of the central nervous system that is an obligatory intermediary between the brain and most of the body. From the neck down, physiological sensory information and nearly all motor and visceral effector signals are trafficked through neurons whose cell bodies and nuclei sit in the spinal cord. If a larger body required more neurons to be operated, the scaling of the number of neurons in the spinal cord should tell us how much of a demand was imposed.

So far, we only had data on the neuronal composition of the spinal cord in primates, but those were very interesting data. We found that, even though the mass of the primate spinal cord scaled with body mass raised to the power of +0.73 across primate species (agreeing with the expectation that brain mass was limited metabolically and thus scaled with body mass raised to a power close to +0.75), the number of neurons in the spinal cord scaled with body mass raised to the much smaller power of +0.36: a

1,000-fold heavier primate only had about 10 times more neurons in its spinal cord. This was far from the expected correlation with body mass (which would yield the scaling exponent +1.0), with body surface (where the exponent would be +⅔ or +0.67), or with metabolic rate (where the exponent would be +0.75). Rather, the observed exponent was close to the scaling of the linear dimension of the body (body length) with body mass raised to the power of +⅓ or +0.33 (figure 8.2). The number of neurons in the primate spinal cord thus seemed to scale simply with body length,[5] and we confirmed that it indeed grew proportionally with the length of the spinal cord across primate species, with, on average, 43,000 neurons per millimeter—which is very few neurons indeed. We now know that the mouse spinal cord has about 2 million neurons,[6] and the human spinal cord, large as it is, has an estimated 20 million neurons, just ten times the number in a mouse spinal cord and only one-third the number of neurons found in an entire mouse brain.[7] Put this way, it seems remarkable that we

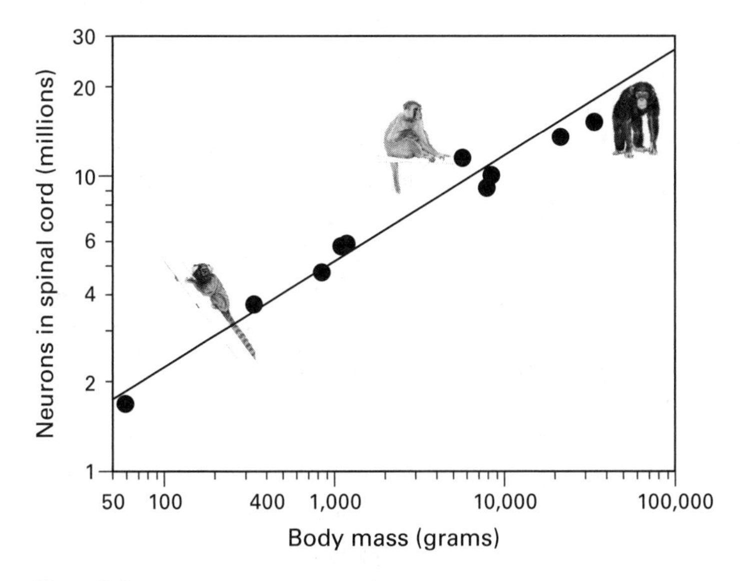

Figure 8.2
Number of neurons in the spinal cord scales with body mass raised to the power of +0.36 across primate species. The number of neurons expected for the chimpanzee spinal cord is indicated on the same graph. Whereas body mass of the species examined varies by a factor of nearly 1,000, the number of neurons in the spinal cord varies by a factor of only 10.

have so much control over our bodies with so few neurons in the spinal cord—and it becomes understandable that even small lesions to the spinal cord can cause catastrophic damage to that control.

What does the slow scaling of the number of neurons in the spinal cord mean to the brain-body relationship? The very small scaling exponent (+0.36) means that we can safely discard the previous hypotheses that the growing body mass or body surface imposes a strong need for more neurons. Likewise, we can discard the hypothesis that the number of neurons in the spinal cord is limited by the scaling of the metabolic rate of the body, for it does not increase fast enough for energy to be a problem. Instead, it seems that the number of neurons in the spinal cord is related simply to its length. Actually, given that the length of the spinal cord must be (in developmental terms) rather a *result* and not a cause of its number of neurons, we have proposed that (1) factors yet to be determined control how many neurons go into the developing and then adult spinal cord; (2) at a distribution of 43,000 neurons per millimeter, the number of neurons determines, still early in development, how long the adult spinal cord will be; and (3) the body then grows around it, related to, but not constrained by, that cord length and its number of neurons.[8]

Because the spinal cord contains both sensory and motor neurons, it was still possible that the number of motor neurons, a small subset of all spinal cord neurons, scaled linearly with muscle mass. One way to examine that possibility was to count motor neurons directly. This was done by my colleague Charles Watson in the facial motor nucleus of marsupial species and independently by Chet Sherwood in the facial motor nucleus of primates[9] (the facial motor nucleus, incidentally, sits in the rest of brain, not in the spinal cord). In both cases, the number of motor neurons that control facial movements scaled only very slowly with body mass, which could be used as a proxy for the muscular mass in the head. The scaling exponents here were actually even smaller than what we found for all neurons in the spinal cord: +0.13 across primates and +0.18 across marsupials[10] (figure 8.3). Remarkably, just a few thousand motor neurons appeared to be enough to control the entirety of facial movements in small and large marsupials and primates alike—even as body mass, and with it the mass of facial muscles, scaled by a factor of more than 10,000.

With the scaling exponent +0.18, doubling the body mass is accompanied by a modest increase of only 14 percent in the number of facial

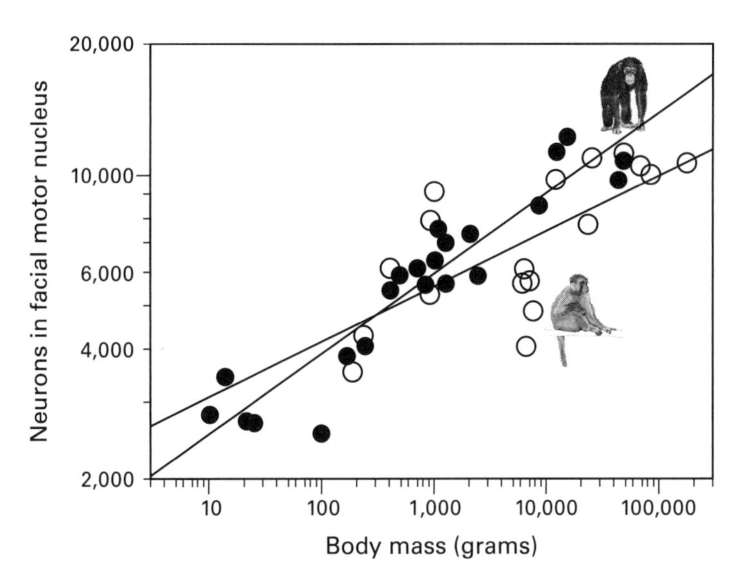

Figure 8.3
Number of motor neurons in the brain's facial motor nucleus, which controls the movements of the face, scales with body mass raised to the power of +0.18 across marsupials (filled circles) and the power of +0.13 across primate species (open circles). For a similar body mass, the facial movements of marsupials and primates are controlled by similar numbers of motor neurons—and quite few of them.

motor neurons, and halving the body mass is accompanied by a decrease of only 12 percent. Interestingly, these numbers are remarkably similar to the seminal findings of Viktor Hamburger and others in the 1970s[11] on how numbers of motor neurons in the spinal cord are adjusted during development, through differential cell death and survival, to the size of the pool of muscle fibers that they innervate. After experimentally doubling the target muscle field of developing birds and amphibians by transplanting an extra limb bud to the developing embryo, or halving the target muscle field by forcing two nerves to innervate the same muscle, those researchers found that the number of spinal cord motor neurons, instead of doubling, increased by only 15–20 percent and, instead of halving, decreased by 8 percent. There *was* some sort of quantitative matching between numbers of motor neurons and the muscle mass to be innervated since more neurons survived when there were more muscle fibers to be innervated, and fewer neurons remained when the muscle pool was decreased.[12] But it remained to be determined why the matching was not linear: why did the experimental doubling of the target muscle field lead not to a similar doubling of the number of motor neurons, but only to an increase of 15–20 percent? Our findings of a similar scaling between numbers of facial motor neurons and a proxy for muscular mass across adult primate species suggest that there is a fundamental competition-for-survival mechanism at work that ties numbers of motor neurons to the mass of muscle fibers to be innervated, leading to a nonlinear numerical matching governed by a power law with a small scaling exponent, both within and across species. Furthermore, if the same competition mechanism operates at both developmental and evolutionary levels, matching numbers of motor neurons to muscle mass, then it is not so much that a larger body *requires* a larger number of motor neurons to control but, rather, that it *allows* a larger number of neurons to survive.* A new picture thus emerges, one in which the central nervous system is formed first, with numbers of neurons that are determined genetically and proportional to the length of the body, but that can be further reduced depending on the actual mass of muscles and sensory targets available for innervation.

What about nonprimate mammalian groups? Although we don't yet have data on the numbers of neurons in the spinal cord of nonprimate species, we do know that the number of neurons in the rest of brain scales linearly with the number of neurons in the spinal cord across primate species.[13] That is, for ten times more neurons in the spinal cord, there are also around ten times more neurons in the primate rest of brain. This linearity supports the notion (1) that both the spinal cord and the rest of brain contain neurons that are directly related to managing the body and (2) that similar mechanisms exist that control the allocation of neurons to these structures, at least in primates. We can look at the scaling of the number of

*The slow scaling of the number of motor neurons with increasing body mass suggests that, as animals become larger in evolution, the number of muscle fibers that are controlled by individual motor neurons (and thus the size of the average motor unit) increases. The expected consequence of such an increase is that motor control should become less and less precise and refined in larger species—but, at least in primates, this might be offset by a faster scaling of numbers of cortical motor neurons relative to spinal motor neurons (Herculano-Houzel, Kaas, and de Oliveira-Souza, 2015).

neurons in the rest of brain across a larger number of nonprimate species to gain insight on their relationship to body mass.

Again we find that different relationships apply to primates and nonprimate species: as shown in figure 8.4, the rest of brain has more neurons in primates than in nonprimate mammals of a similar body mass, and this number scales faster across primates (with body mass raised to the power of +0.5) than nonprimates (with body mass raised to the power of +0.3). And again, these scaling exponents are both much smaller than the exponent +1.0 expected if a larger muscle mass required a proportionally larger number of neurons and smaller than the exponent +0.75 expected if metabolism imposed a limitation on the number of neurons. If larger bodies do indeed require more neurons to operate them, they do so at a very slow pace.

Remarkably, the human rest of brain contains about as many neurons as would be expected for a generic primate of our body mass. There is no

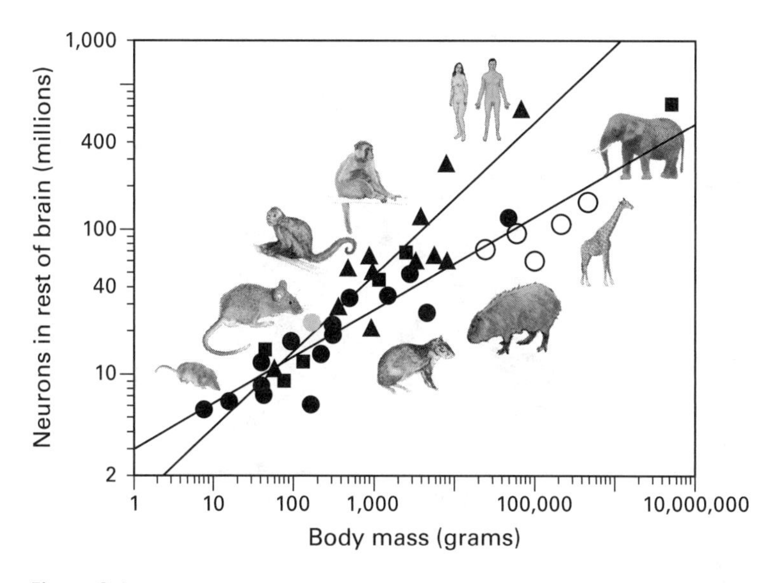

Figure 8.4
Number of neurons in the rest of brain scales with body mass raised to the power of +0.5 across primate species (triangles), but to the smaller power of +0.3 across all other species (afrotherians, squares; eulipotyphlans and rodents, filled circles; artiodactyls, open circles). For a similar body mass, primates have more neurons in the rest of brain structures than nonprimates. The number of neurons in the human rest of brain matches the number expected for a generic primate of its body mass.

extraordinarily large number of neurons in the human rest of brain to deal with the body: our brain has the mass and the number of neurons in the rest of brain that a non–great ape primate brain of our body mass would be expected to have.

Primates also have more neurons in both the cerebral cortex (figure 8.5) and the cerebellum than nonprimates of similar body mass (figure 8.6). There is, therefore, no single way to scale the number of neurons in a mammalian brain to the corresponding body mass—and even the general scaling of brain mass with body mass hides the fact that primates have many more neurons in their brains than nonprimate mammals of similar body mass have because of the different neuronal scaling rules that apply to their brains. The logical implication is that the number of neurons in the mammalian brain is not *determined* by the size of the body—even if other factors still lead to a correlation. We should consider the inverse possibility, however: that the increasing number of neurons in the brain might

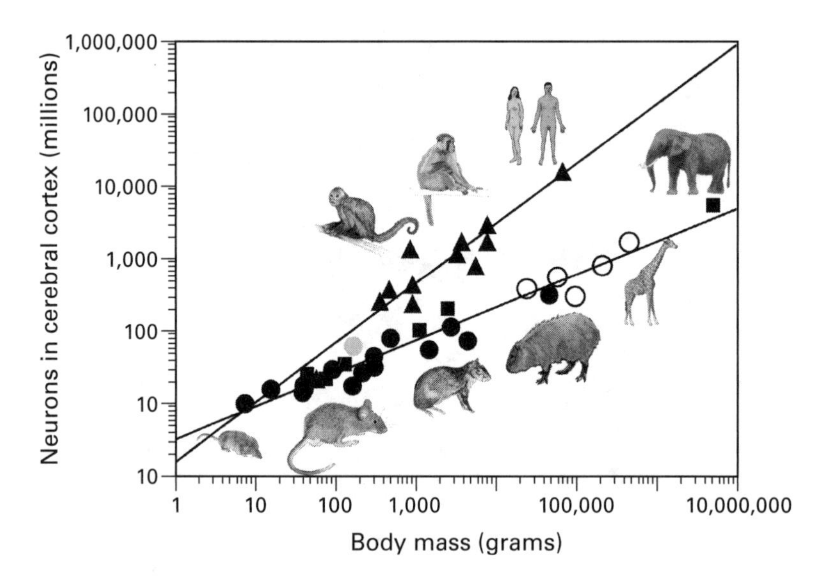

Figure 8.5
Number of neurons in the cerebral cortex scales with body mass raised to the power of +0.8 across primate species (triangles), but to the smaller power of +0.5 across all other species (afrotherians, squares; eulipotyphlans and rodents, filled circles; artiodactyls, open circles). For a similar body mass, primates have more neurons in the cerebral cortex than nonprimates.

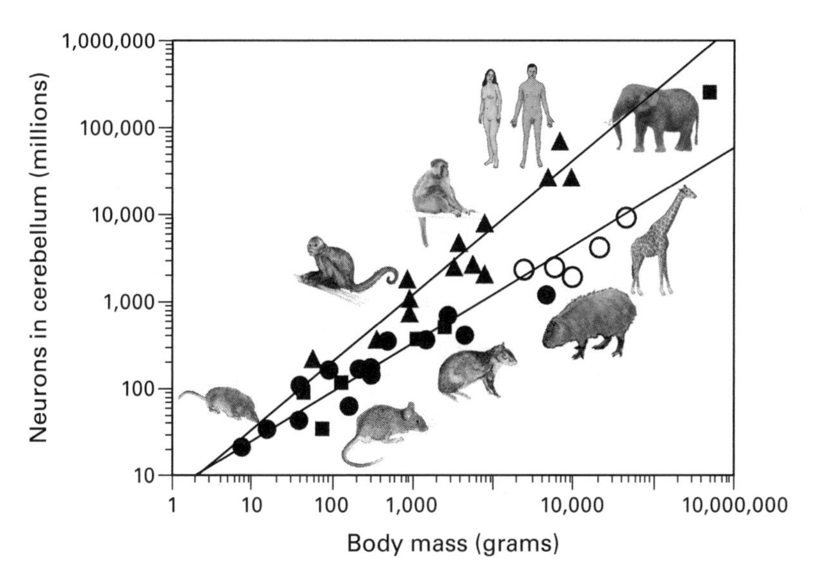

Figure 8.6
Number of neurons in the cerebellum scales with body mass raised to the power of +0.8 across primate species (triangles), but to the smaller power of +0.5 across all other species, not including the elephant, which has an extraordinarily large number of neurons in the cerebellum for its body mass (afrotherians, squares; eulipotyphlans and rodents, filled circles; artiodactyls, open circles). For a similar body mass, primates have more neurons in the cerebellum than nonprimates.

somehow be facilitated by a larger body mass in mammalian evolution, as I proposed recently.[14] This would be especially important for small animals, such as the first mammals, because increasing body mass would have allowed them to obtain more energy per hour of foraging and feeding. But that is a whole other story.

In any case, primates have yet another advantage over nonprimate mammals: they have many more neurons in their cerebral cortex, cerebellum, and rest of brain than nonprimate mammals of similar body mass have. Given the emphasis that for so long was placed on the size of the body and how it supposedly determined the number of neurons in the brain, primates have long been underestimated. Indeed, expecting the body-brain relationship that applies to nonprimate mammals to extend to primates is like expecting apples to be oranges on the inside: a primate brain is unlike the brain of any other mammal when it comes to how many neurons fit in it, how many neurons operate the body that goes with it, and how many functions it performs that go way beyond simply moving that body.

Yet again, humans are no exception once they are compared to their kin: other primates (great apes so far excluded simply for lack of data). As figures 8.5 and 8.6 show, our cerebral cortex and cerebellum are the largest among primates, with the most neurons—but, once more, these structures hold just the number of neurons expected for a generic primate of our brain and body mass. Everywhere we look, the same pattern is confirmed again and again: humans are not special. We are a large primate species, with the large number of brain neurons that comes with it. So much for encephalization, then, since our brain is not too large for our body mass. Rather, it is the brains of great apes that are too small for their body mass.

That is made clear by applying the scaling rules shown in figures 8.4–8.6, which predict that orangutans and gorillas, whose males weigh, on average, 70 kilograms (155 pounds) and 125 kilograms (275 pounds), have bodies that are way too large for their numbers of brain neurons: generic primates with as many cells in their cerebellums as gorillas have should weigh only about 24 kilograms[15] (53 pounds), whereas an actual gorilla can weigh up to 275 kilograms (605 pounds) or almost twelve times as much. How did great apes, which have even larger bodies than we do, come to have so much smaller a number of neurons in their brains than expected?

Much more fundamental, however, is what the discrepancy between great apes and other primates means at heart: that the relationship between body mass and the number of brain neurons is not a very tight one. Even though there is, in general, a good correlation between body mass and the number of brain neurons, there is evidently also a good deal of flexibility in how large a body can grow around a brain with a certain number of neurons—and great apes are proof of that flexibility.

Do Larger Bodies Come with Larger Neurons?

It seems intuitive and likely that animals with larger bodies have larger numbers of cells. Although popular estimates of the number of cells in the human body range around a few trillion, there have been no real estimates of how many cells compose the human body, just as estimates of the number of neurons in the human brain were for a long time simply order-of-magnitude extrapolations, rather than real estimates. At best, the average density of cells measured in a few bodily tissues and organs can be multiplied, first, by the total volume of the tissues and organs and, then, by the reciprocal of the fraction of body volume the tissue or organ represents to provide, once again, an order-of-magnitude extrapolation for the total number of cells in the body.

Estimates of cell density in different organs do, however, paint an interesting picture of a related aspect of what happens as bodies become larger—or rather, of what makes a body become larger. Cells in larger bodies become more numerous, yes, but they also seem to become larger as well. At least, that appears to be the case for the liver and skin.[16]

If the brain behaves as just another body part, then larger brains—that is, brains with more neurons—should also have larger cells. As we saw in chapter 4, while this pattern does apply to the brains of nonprimates, it stopped applying to the cerebral cortex and cerebellum of primates, which gain neurons without the average size of those neurons becoming much larger.

The rest of brain of primates has also diverged away from the ancestral pattern, with smaller increases in average neuronal cell size than expected. Across nonprimate species, neuronal density in the rest of brain decreases steeply as it gains more neurons, indicating that average neuronal size increases jointly with numbers of neurons in the rest of brain. In primates, however—and, again, humans are no exception—there is no significant decrease in neuronal density as the rest of brain gains neurons (figure 8.7). That is, if average neuronal size becomes larger, it does so much more *slowly* than in nonprimate mammals. For similar numbers of neurons in the rest of brain, neurons in these noncortical, noncerebellar structures are much smaller in the brains of primates than in those of nonprimates. Because average neuronal size is inversely proportional to neuronal density,[17] we can estimate by how many times average neuronal size varies across species with similar numbers of neurons. The kudu (an artiodactyl) and the rhesus monkey (a primate) have between 105 and 125 million neurons in the rest of brain—but, given their different neuronal densities, we estimate that the average neuron in the kudu's rest of brain is almost 10 times as large as in the macaque's. As a consequence, the kudu's rest of brain (at 65 grams) is more than seven times the mass of the rhesus monkey's (at 9 grams).

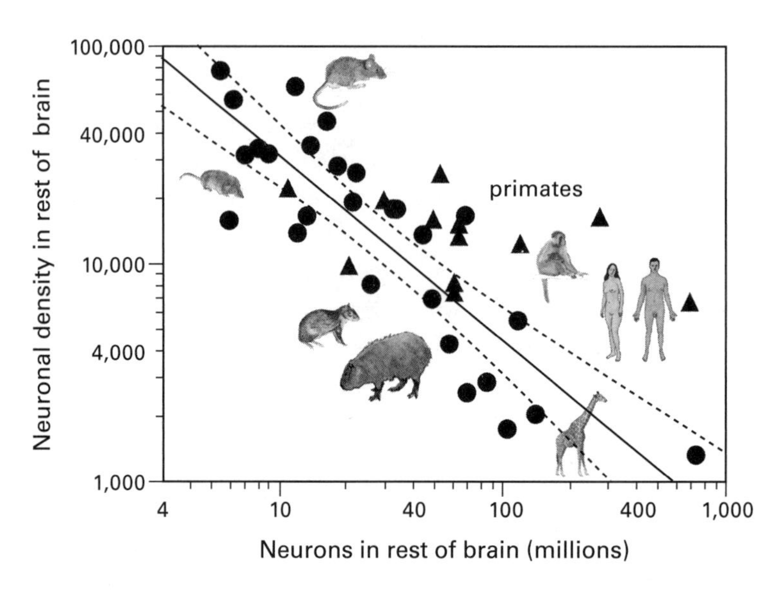

Figure 8.7
There is a strong, negative correlation between neuronal density (in neurons per milligram of rest of brain) and number of neurons in the rest of brain across nonprimate species (circles); density scales with number of neurons to the power of –0.9, which implies that the average mass of neurons in the rest of brain increases with the number of neurons raised to the power of +0.9. In contrast, neuronal density does not decrease significantly across primates (triangles), as the rest of brain gains neurons. The strikingly different neuronal densities across primates and nonprimates with similar numbers of neurons in the rest of brain mean that its neurons are much smaller in primate than in nonprimate species.

The discrepancy between primate and nonprimate mammals in how neuronal size in the rest of brain increases as it gains neurons indicates a clear evolutionary break of the primate lineage from its common ancestor with nonprimate mammalian lineages. But due to what? Here's one interesting possibility: because primates are much smaller than other terrestrial mammals of similar brain mass, such as artiodactyls (cows, antelopes, giraffes) and the largest rodents, could it be that their smaller average neuronal size in the rest of brain is related to their smaller body mass, which, in turn, drove their rest of brain scaling rules to change toward neurons that became larger more slowly as they became more numerous?

The analysis shown in figure 8.8 indicates that a smaller body mass compared to that of nonprimate mammals can indeed account for the smaller neuronal cell sizes in the rest of brain of primates. Figure 8.8 shows that the density of neurons in the rest of brain decreases across *all* mammalian species examined—primates, artiodactyls, and others—as a function of body mass, with the exponent –0.3. We can infer, therefore, that neurons in the rest of brain became, on average, larger as the body gained mass, and they did so across all mammalian species examined, including primates. Primates maintained the same scaling of neuronal size in the rest of brain with increasing body mass as nonprimate mammals, even as they diverged away from nonprimates with a larger number of neurons in the rest of brain for their body mass. This difference, by the way, indicates that even though the average size of neurons in the rest of brain is related to body mass, the number of neurons in the rest of brain is not.

It makes sense that average neuronal cell size increases in the rest of brain as a function of increasing body mass. The "rest of brain" includes the

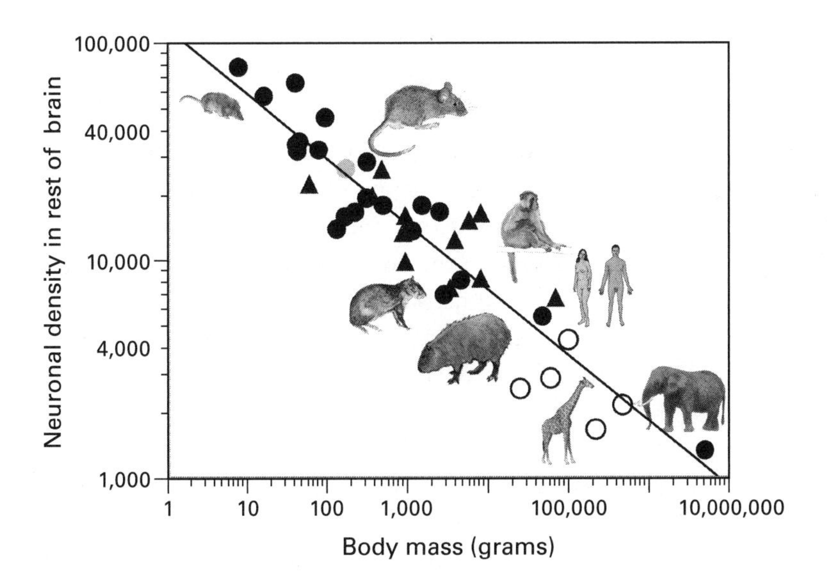

Figure 8.8
Very diverse mammalian species (artiodactyls, open circles; eulipotyphlans, afrotherians, and rodents, filled circles; primates, triangles) share a single, inverse relationship between body mass and neuronal density in the rest of brain (in neurons per milligram), which indicates that neurons in the rest of brain on average become larger with increasing body mass across all species alike. This power function has the exponent –0.30.

medulla and pons, whose sensory neurons collect information directly from the body, and whose effector neurons (both motor and visceral) manage bodily functions directly. If these neurons are to remain connected to their targets in the body, they must have fiber connections that become elongated as the body elongates. Fiber length is expected to scale with the linear dimension of body volume, that is, with body volume or mass raised to the power of +⅓ or +0.33. If all neurons in the rest of brain scaled similarly, becoming elongated at the same rate as the body increased in length, their average neuronal mass would be expected to scale with body mass raised to the power of +0.33—and neuronal density would be expected to scale in the opposite direction, with body mass raised to the power of –0.33, very close to the scaling exponent of –0.3 observed. Average neuronal mass in the rest of brain thus scales with the linear dimension of the body, and the increase in body length is a very likely candidate mechanism to drive this increase in average neuronal cell mass by elongation of their fibers contacting targets in the growing body.

Could a similar mechanism apply to the cerebral cortex and cerebellum? Are variations in neuronal density in these structures, and therefore average neuronal cell size, also tied directly to variations in body mass? Figures 8.9 and 8.10 show that this is not the case. Neuronal density in the cerebral cortex does indeed vary as a power function of body mass with a similar scaling exponent, –0.29, suggesting that average neuronal cell size also scales with the linear dimension of the body, but only across nonprimate species; across primates, neuronal density in the cerebral cortex does not vary significantly as a function of body mass. Thus body mass can't be a direct determinant of average neuronal cell size in the cortex as it is for the rest of brain, or the same function would have to apply across primates and nonprimates alike.

In the cerebellum, neuronal density also varies as a power function of body mass, but with a smaller exponent, –0.16, and only across nonprimate, noneulipotyphlan species (figure 8.10). Across the latter, there is no significant relationship between body mass and neuronal density in the cerebellum, just as there was no relationship between neuronal density and number of cerebellar neurons. The different scaling of cerebellar neuronal density with body mass across mammalian groups suggests that body mass is also not a direct determinant of average neuronal cell size in the cerebellum, or the same function would have to apply across all species alike.

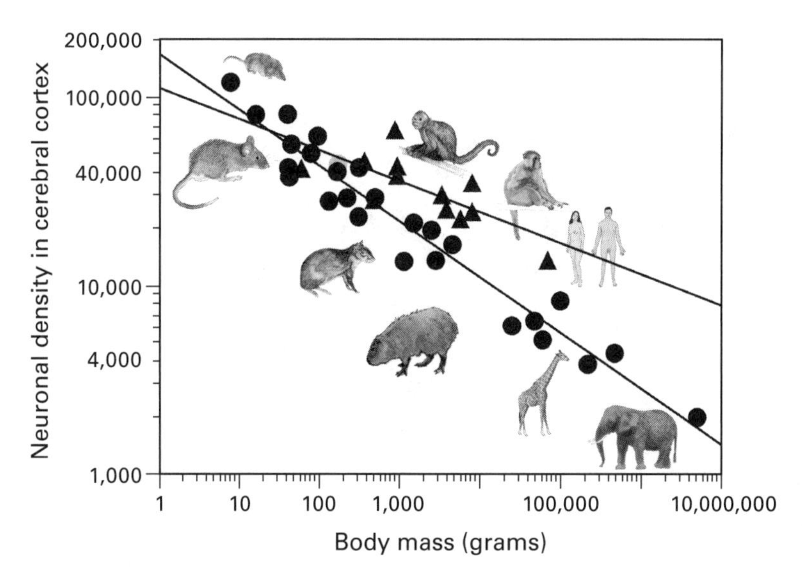

Figure 8.9
Neuronal density in the cerebral cortex (in neurons per milligram) scales with body mass across nonprimates (circles) but not across primates (triangles). Neurons in the cerebral cortex on average become larger as a power function of increasing body mass with the exponent –0.29 across nonprimates.

We can conclude, then, that larger bodies across mammalian species (humans included) do appear to come universally with larger neurons in the rest of brain—and this is probably a consequence of the very increase in body mass, which necessarily leads to longer, and therefore larger, neurons in the rest of brain structures connected directly to the body. But in the cerebral cortex, which is removed from direct contact with the body, and which instead receives information from the thalamus (in the rest of brain), neurons do not become uniformly larger in larger animals—although they do tend to increase in average mass together with body mass, even across primates. In the cerebellum, average neuronal size may or may not increase together with body mass across species. And even if the average mass of neurons in the rest of brain scales universally with increasing body mass, the number of neurons in the rest of brain is free to scale differently across primates and nonprimates. There is no one single way to add neurons to larger bodies. Evolution has allowed, at the same time, both concerted changes in some aspects of some structures, and different, independent

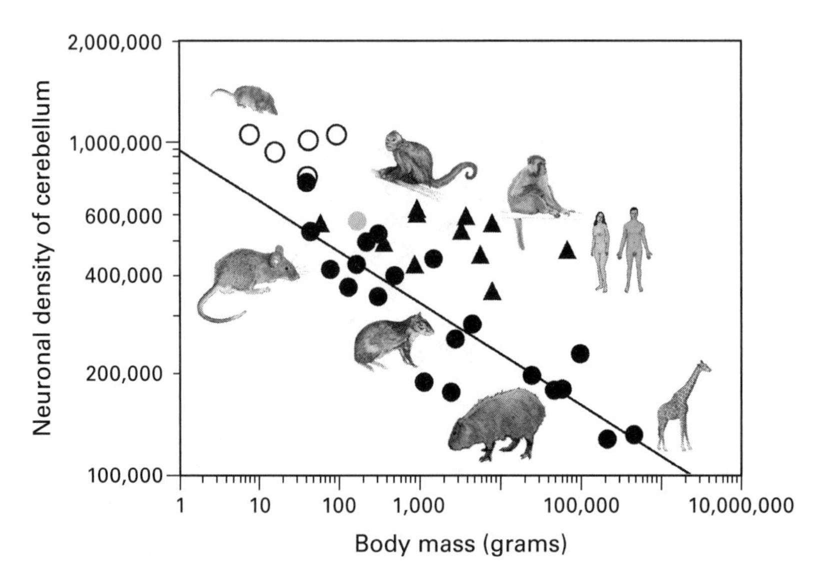

Figure 8.10
Neuronal density in the cerebellum (in neurons per milligram) also scales differently with body mass across primates (triangles), eulipotyphlans (open circles), and other species (filled circles). Neuronal density in the cerebellum of noneulipotyphlan, nonprimate species scales as a power function of increasing body mass with the exponent –0.16, but does not scale significantly in size across eulipotyphlans or primates of increasing body mass (as indicated by the lack of significant correlation between neuronal density across these species and body mass).

changes in others—which is the basis of the concept of "mosaic" evolution: that different parts of the brain are free to change in different directions as they evolve.

And if there isn't a single way to add neurons to larger bodies, then Jerison's calculation of an encephalization quotient for humans by comparison to all mammals, primates and nonprimates alike, is based on a faulty assumption—that humans are comparable to *all* other mammals. As we have seen, humans can properly be compared only to other non–great ape primates. How great apes came to be the clear exceptions to the body-brain relationship and why that matters is the subject of the next two chapters.

9 So How Much Does It Cost?

As it turns out, then, the human brain is not special in its number of neurons: we have just as many neurons as expected in the cerebral cortex, cerebellum, and rest of brain for a non–great ape primate of our body and brain mass. Our cerebral cortex is relatively large in mass, compared to the rest of brain, but has the same 15–25 percent of all brain neurons as the cortices of other mammals (African elephant aside, with its enormous neuron-rich cerebellum), and our prefrontal cortex holds 8 percent of all cortical neurons, as in other primates. Our single largest advantage over other animals is most easily ascribed to the sheer number of neurons available in the cerebral cortex, and in the prefrontal cortex in particular, to process information in complex, flexible ways that can predict future outcomes and act as required, particularly in a way that intelligently maximizes future possibilities.

On the other hand, the human brain does seem special in the amount of energy that it requires. As mentioned in chapter 1, even though it represents only around 2 percent of body mass, the human brain alone costs some 500 kilocalories per day to operate, a disproportional 25 percent of the daily energy required for the entire human body to work.[1] In comparison, the brains of other vertebrate species cost at most 10 percent of the daily energy budget of the body.[2] The remarkably higher relative energy cost of the human brain certainly makes it seem special. The source of all this energy is usually glucose from the blood, which crosses the blood-brain barrier, may or may not be broken down first into lactate by glial cells, and is shuttled to neurons as required according to their activity.[3] This explains why blood flow to the brain is tightly regulated: to keep the human brain running on blood glucose requires a remarkably constant average of 750 milliliters of blood flowing through the brain per minute, and even a 1 percent drop in flow may cause fainting and loss of consciousness.

It's no surprise to find that the brain is relatively expensive, given that it is the second most energy-costly organ in the body, ranking in total organ metabolic cost below the liver only.[4] But even though the metabolic needs of most body organs are closely associated with body size, such that the relative metabolic cost of an organ depends on its relative size compared to the body,[5] the relative metabolic needs of mammalian brains are variable. A macaque's brain uses 13 percent of the energy needed to run its body; a shrew's brain, only 1 percent of its body's daily energy expenditure.[6]

And then there is the human brain, which uses around 25 percent of the energy it takes to run the entire body. This is all the more puzzling because its "mass-specific" metabolic cost—its energy cost per gram of brain tissue per minute, at 0.31 micromole (μmol) of glucose—is only about *one-third* the energy cost per gram of mouse brain tissue per minute, at 0.89 micromoles.[7] This means that one gram of human brain is three times cheaper to run than one gram of mouse brain in terms of energy, which is counterintuitive, to say the least: with its prowess, shouldn't a human brain consume *more* energy per gram of tissue than a mouse brain? To add to the puzzle, the low mass-specific metabolic cost of the human brain also seems at odds with evidence of the genetic upregulation of energy metabolism in human evolution.[8]

So many contradictions had one clear meaning to me: we simply didn't understand how brain metabolism scaled across brain sizes in evolution. If the human brain was not an outlier in its number of neurons, weighed only 2 percent of body mass, and had a low mass-specific metabolism, how did it come to cost so much more of the energy the human body needed to function?

The Cost of Running a Brain

Intuitively, larger brains should cost more energy, just as larger bodies do, with a metabolic cost that has been known since Max Kleiber's studies in the 1930s[9] to scale with the mass of the body raised to the power of +0.75. The exponent +0.75 tells us that a 10 times larger body costs 5.6 times more energy per day, and a body 100 times larger costs 31.6 times more. The larger overall energy requirement of larger bodies makes sense: keeping cells alive and well costs energy, and the more cellular stuff there is to keep organized and away from the entropic equilibrium that is death, the more energy they require. But why and how exactly the amount of energy required scales with the cellular or body volume not linearly, but with the smaller-than-unity allometric exponent +0.75 remains one of biology's greatest mysteries.

By the end of the twentieth century, it was widely believed that larger brains were universally made of larger neurons and that larger neurons cost more energy.[10] If glial cells were the providers of energy to neurons, then larger neurons should require a larger proportion of glial cells per neuron to keep them running, as Andrew Hawkins and Jerzy Olszewski had proposed in 1957.[11] Interestingly, however, the notion that glial cells exist in a numerical proportion to neurons related to their metabolic maintenance preceded any experimental evidence both on how much energy neurons cost and on whether glial cells actually provided metabolic support to neurons. But the notions that larger neurons required more energy and that a larger proportion of glial cells per neurons provided that energy seemed, for the longest time, to make perfect sense—until the actual numbers turned up.

Glial Cells to Keep Neurons Running

Often called the "other" type of cell in the brain and named as second best to neurons, glial cells, or glia, got their name first in German, as "Nervenkitt" (nerve glue), from neuroanatomist Rudolf Virchow in 1856, and later in English from the Greek word for "glue," to describe the role they were once supposed to play, to fill in the gap between neurons. For decades in the twentieth century, and based on hardly any data, glial cells were widely believed to be the most common cells in the brain, occurring in numbers that increased with brain mass faster than the numbers of neurons themselves—to the point that glial cells supposedly outnumbered neurons by a factor of 10 in the human brain. In that sense, the ratio lent credence to the popular notion that we used only 10 percent of our brain cells—the neurons. Science journalists loved the round numbers, using catchy opening statements such as "The most numerous type of cell in the human brain—outnumbering neurons ten to one"[12] or "Meet the forgotten 90 percent of your brain: glial cells, which outnumber your neurons ten to one."[13] The journalists are not to blame, though: It said so, after all, in Eric Kandel's *Principles of Neural Science*,[14] as well as in Mark Bear's *Neuroscience: Exploring the Brain*.[15] A number of the best specialists in glial cell biology also endorsed such statements in their own scientific papers,[16] even as they provided evidence of the crucial role that glial cells are now known to actually play in brain tissue. A larger number of glial than neuronal cells seemed to make their role in the brain all the more important.

Glial cells have indeed ascended from their role as mere supporting actors to the status of key players in brain physiology, metabolism, development, and even disease:[17] they control synapse formation and function, regulate synaptic transmission, respond to neural activity, and are, in fact, metabolically coupled to neurons, providing them and their axons with lactate as a source of energy on demand.[18] Such important functions seemed more in agreement with a cell type that constituted far and away the largest portion of brain tissue. If only that were true.

A Bit of History First

Myths don't materialize out of thin air, and indeed there was some evidence in the early days of neuroscience that something interesting was going on with the proportion of glial to neuronal cells. It started with the German neuropathologist Franz Nissl who, still in 1898, inspected sections of mole, dog, and human brains under his microscope and concluded that neuronal density (neurons per volume) decreases as the volume of the cerebral cortex increases, with neuronal density lowest in the human cortex. Nissl attributed the decrease in neuronal density across species not to an increase in neuronal cell size, but rather to an increase in the "nonneuronal portion of the tissue"—which was, in his view, evidence of a higher development of "psychic functions" in humans.[19]

A half century later, in 1954, the German neuroanatomist Reinhard Friede compared the cerebral cortex of a number of species and observed that the ratio between the number of glial cells and the number of neuronal cells (which later became known as the "glia/neuron ratio" in the literature) increased from frog (0.25) to human (1.48, on average, across cortical layers), going in ascending order of brain size from mouse (0.36), to rabbit (0.43), to pig (1.20), to cow (1.22), to horse (1.23).[20] Friede endorsed Nissl's conclusion, well in line with Edinger's view of progressive brain evolution:

the "progressive development" of the cortex was associated with a relative increase in the glia/neuron ratio—with humans as the "most developed," of course. Such a relative increase in numbers of glial cells was an indication of their "trophic importance," given the suspected involvement of glial cells in brain metabolism, thus allowing the presumed "progressive development" to occur. Having more glial cells per neuron than the brains of other species might be why the human brain could achieve more with its neurons.

But that was only because the human brain had, until then, been the largest one to be analyzed. In 1957, using tissue from fin whale brains (each weighing around 7 kilograms) that had been examined by Donald Tower and Allan Elliott,[21] Hawkins and Olszewski[22] found a much higher glia/neuron ratio in the whale cerebral cortex, 4.54, compared to 1.78 in the cortex of a human brain weighing only 1.5 kilograms. So much for "progressive development": the glia/neuron ratio might simply reflect brain size, as was confirmed by Herbert Haug in a seemingly comprehensive meta-analysis of dozens of species,[23] illustrated in figure 9.1. Hawkins and Olszewski proposed that the increase in the glia/neuron ratio was related to an increase in the size of the neurons, which, with longer processes, would "require more assistance from the support tissue to cater to their metabolic needs." It all seemed to fit: Donald Tower,[24] looking at a number of mammalian species lumped together, had shown that larger brains had lower neuronal densities, which suggested larger neurons;* larger neurons should, intuitively, have greater metabolic needs, requiring support from more glial cells; hence, larger brains should have higher glia/neuron ratios.

There was reasonable evidence of an increasing proportion of glial cells over neurons in larger brains, as seen in figure 9.1. But all these data were for the cerebral cortex alone—and the human cortex did not appear in any way special: there was no evidence of anything like a 10 to 1 ratio of glial to neuronal cells in the human cerebral cortex, much less in the whole human brain. That particular ratio seems to have been passed along from scientist to scientist in a neuroscientific version of the "telephone" game in the literature, as uncovered by my colleague Christopher von Bartheld.[25]

*As seen in chapter 4, it turns out that only nonprimate species have larger brain neurons.

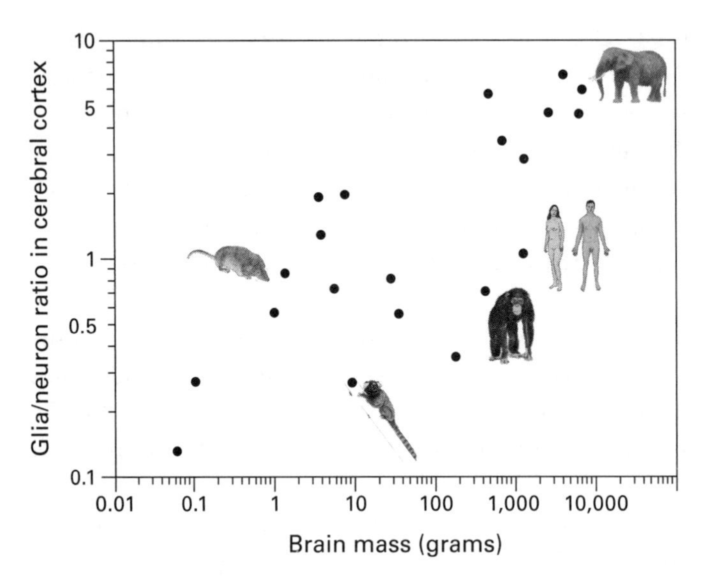

Figure 9.1
Ratio of numbers of glial to neuronal cells in the cerebral cortex seems to increase together with brain size across species as diverse as the marmoset, mole, cat, human, African elephant, and various species of whales. Data from Haug, 1987, Stolzenburg, Reichenbach, and Neumann, 1989, and Hawkins and Olszewski, 1957.

Back to the Future

Now that we had actual data on the numbers of cells that composed different parts of the brain of different species, we could look at what indeed happened with the glia/neuron ratio as the brain increased in mass across species. But, first, a brief disclaimer. Although we have a reliable universal marker for neurons (neuronal nuclear protein, NeuN, expressed only in neuronal cell nuclei), we still don't have one for either of the remaining two groups of brain cells: glial and endothelial cells, both of which are therefore counted together as nonneuronal cells. What I am calling "glial cells" or "glia" in regard to our data thus includes what we expect to be a small minority of endothelial cells, which constitute the brain's capillaries and amount to no more than 4 percent of total brain tissue. As such, the number of "glia" in our graphs represents an upper boundary for the total number of proper glial cells in the brain tissue. That disclaimer made, we can move on to consider what happened to the glia/neuron ratio as brain mass increased.

And what happened is … nothing that could be identified as a universal, systematic trend, as figure 9.2 shows. There was neither the expected universal increase in the glia/neuron ratio with brain mass nor a general numerical predominance of glial cells over neurons. For starters, most of the forty-one species in our data set had more neurons than glial cells in their brains: all eulipotyphlans, most rodents and primates, and even the elephant, owner of the largest brain in our data set. Artiodactyls, with brains about as large as those of midsized primates, but much smaller than the elephant brain, were the only animals with consistently more glial cells than neurons in the brain. Even the capybara, with a brain mass of only 75 grams, had a higher glia/neuron ratio in its brain than the elephant. There was no trend for larger brains to have larger and larger

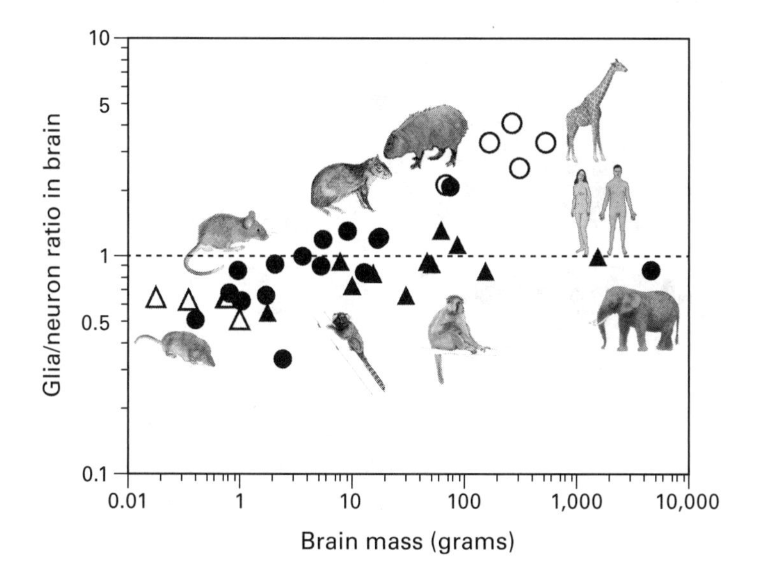

Figure 9.2
Ratio of numbers of glial to neuronal cells in the whole brain has no obvious overall correlation with brain size across species: it increases together with brain mass in rodents (filled circles), but not across eulipotyphlans (open triangles) or primates (filled triangles). The dashed line indicates a ratio of 1, below which neurons are more numerous than glial cells. Only artiodactyls (open circles) have brains consistently composed of more neurons than glial cells. Remarkably, the African elephant brain, the largest in our data set, has a glia/neuron ratio of only 0.84—that is, it has more neurons than glial cells—as do the brains of most of the species in our data set.

proportions of glial cells over neurons. And the human brain did not have 10 times more glial cells than neurons: with an average of 86 billion neurons and 85 billion glia cells, the proportion between glial cells and neurons was almost exactly 1 to 1—much the same ratio we found in other primates.[26]

There was also no systematic trend for an increase in the glia/neuron ratio within each brain structure as it increased in mass (figure 9.3). The glia/neuron ratio was extremely low in the cerebellum, where for every glial cell there were from two to as many as ten neurons. In contrast, glial cells typically predominated in the cerebral cortex and rest of brain—and not simply because the data for the cerebral cortex shown in figure 9.3 included

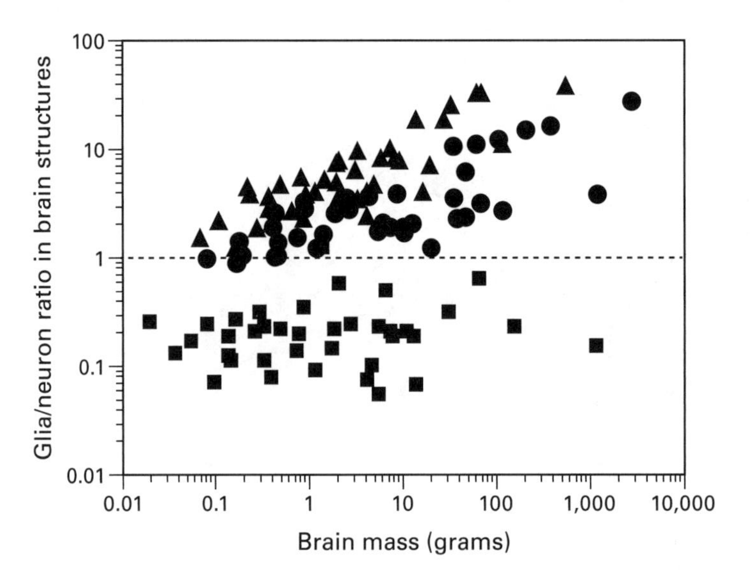

Figure 9.3

There is no universal correlation between the glia/neuron ratio in a brain structure and the mass of the structure across species: it increases in the cerebral cortex (circles) and rest of brain (triangles) in some mammalian orders, but not all, together with structure mass, but not in the cerebellum (squares). The dashed line indicates a ratio of 1, below which neurons are more numerous than glial cells. In the cerebellum, there are always more neurons than glial cells, and usually at least five times as many neurons as glial cells. In the cerebral cortex and rest of brain, in contrast, glia/neuron ratios are almost always above 1, indicating a predominance of glial cells.

the white matter, given that glial cells predominated even in analyses restricted to cortical gray matter.[27]

The reason why there was no universal relationship between the mass of a structure and its glia/neuron ratio is that, as we saw in chapter 4, there was no single, universal relationship between the mass of a brain structure and its number of neurons across all mammalian species we examined. And I can say this because, much to my surprise, there *was* a universal relationship between the mass of a brain structure—*any* brain structure—and its number of nonneuronal ("glial") cells.

It was an extraordinary finding that I made quite by accident while plotting data for an early paper comparing rodent and primate brain structures, when I decided to superimpose in a single figure all the separate graphs I had made for each structure and mammalian order. To my surprise, they all overlapped. It was such a remarkable, near-perfect overlap that I went back to the data tables and checked and rechecked everything to make sure that there were no double entries, no silly mistakes in typing or in copying and pasting numbers from the source data tables. It was real: although the mass of brain structures can scale as any of several different functions of their number of neurons, depending on which brain structure in which mammalian group is examined, as seen in figure 9.4, it scaled as a single predictable function of the number of *nonneuronal cells* in any brain structure, any species, any order—and that included the human brain and its structures, which just blended in with all the brains and brain structures of all the other mammalian species. Give me the size of a mammalian brain structure—*any* structure in *any* species—and I can predict, with good accuracy, how many glial cells it will have. The larger the brain, the more glial cells it will have; and two brains of the same size will have similar numbers of glial cells. As it turned out, all mammalian brains *were* made the same—in regard to how many glial cells build brain tissue.

The universality of the scaling of mammalian brain structure mass with the number of glial cells implies that there is a single biological rule determining how numbers of glial cells are added to brain tissue, regardless of the part of the brain or the mammalian species in question. The scaling rule is a power function with the exponent +1.05, nearly linear, which means that there is no systematic change (or only very, very small changes)

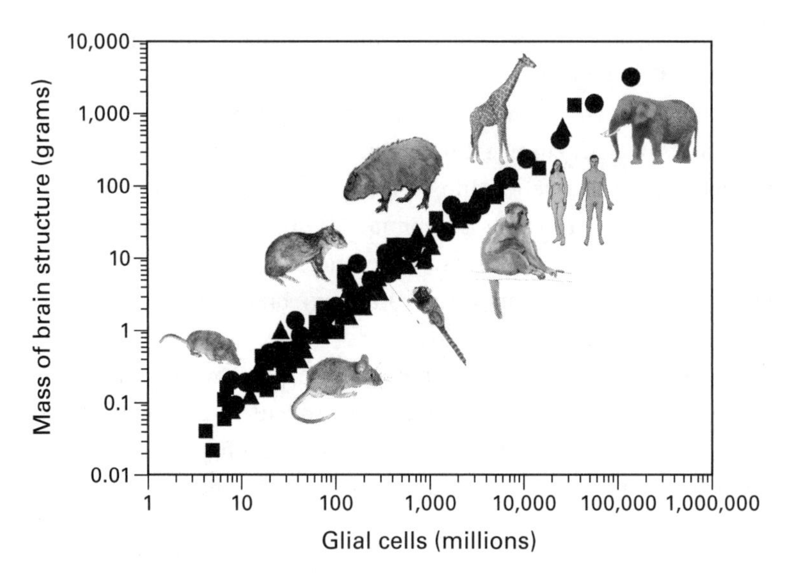

Figure 9.4
There is a universal correlation between the mass of a brain structure and its number of glial cells that applies equally across brain structures, species, and mammalian orders: it is a power function with the exponent +1.05, only slightly above linearity (not plotted).

in the average size of glial cells as they are added to the different brain structures, as indicated by the lack of a trend in glial cell density in figure 9.5: most brain structures had glial cell densities that varied by less than a factor of 3, even as brain mass varied by a factor of 12,500 from mouse to elephant.

If there is a single biological rule governing how numbers of glial cells are added to mammalian brain structures, a rule that is shared by mammalian orders as diverse and divergent in evolution as afrotherians, rodents, primates, eulipotyphlans, and artiodactyls, then it can be safely inferred that that single, shared rule already applied to the ancestor they all have in common—an ancestor estimated to have lived 105 million years ago, as illustrated in figure 9.6. In fact, numbers of glial cells are added to brain tissue in the same way also in birds,[28] which shared a common ancestor with mammals over 300 million years ago. Glial cells must be doing something so sensitive, so important in the brain that the way their numbers are added

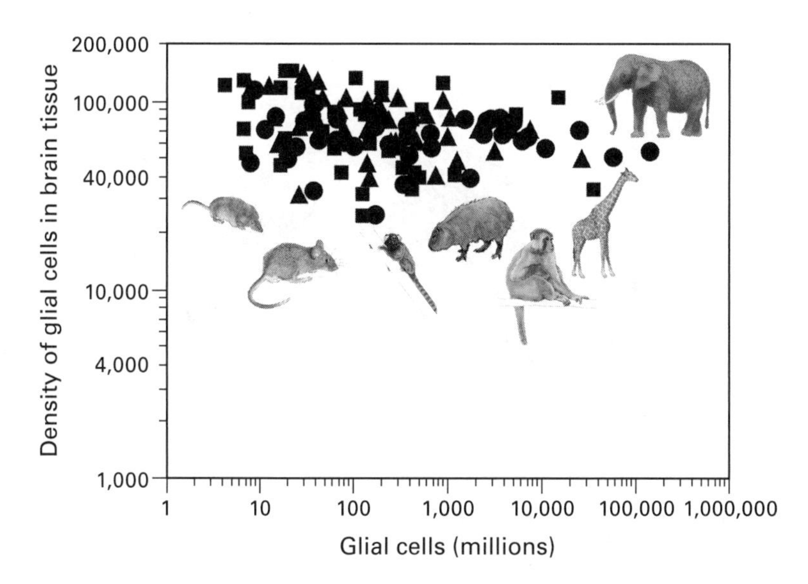

Figure 9.5
Glial cell density varies by a factor of 3 across most brain structures and species, and there is no systematic variation in glial cell density across brain structures as they gain glial cells. This is in stark contrast to the large variation in neuronal density and its systematic decrease in most mammalian orders as brain structures gain neurons. For the sake of comparison, the data plotted here (cerebral cortex, circles; cerebellum, squares; rest of brain, triangles) are shown in the same scale as the plots of neuronal density in chapter 4.

to build brain tissue has remained very much the same over at least the last 300 million years of evolution.

Larger Neurons, Higher Glia/Neuron Ratios

Hawkins and Olszewski had proposed that higher glia/neuron ratios accompanied the larger neuronal size that they assumed, based on Tower's study of neuronal densities, was found in larger brains—well, cerebral cortices, actually, the only structure studied then. We now knew that there was no such universal correlation between larger cortical mass and lower neuronal density, a proxy for larger neuronal cell size. But was the variation in the glia/neuron ratio still related to average neuronal cell size? If so, the glia/neuron ratio in the cerebral cortex should increase

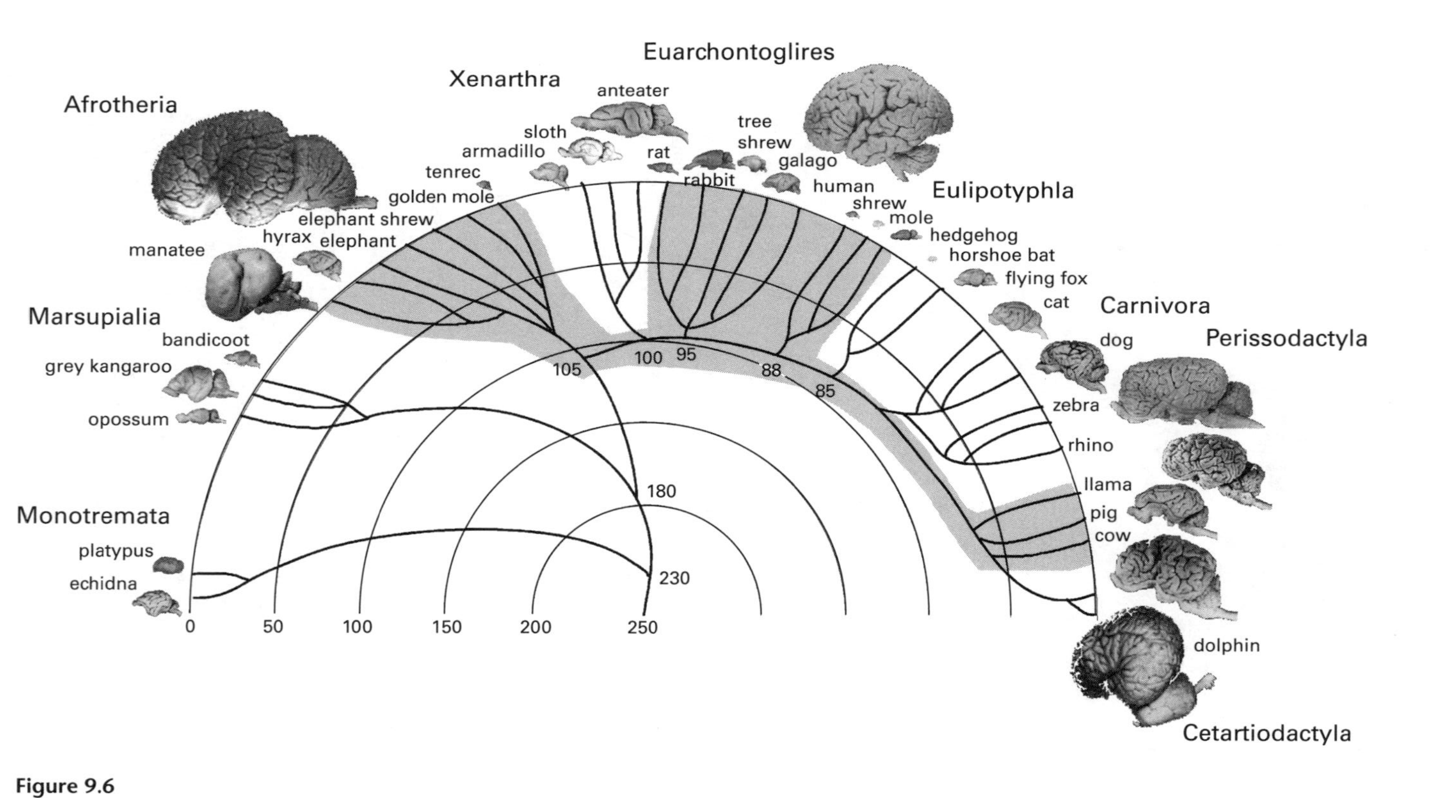

Figure 9.6
Same glial scaling rules apply to groups as evolutionary diverse and distant as afrotherians, rodents, primates, eulipotyphlans, and artiodactyls. Given that their last shared common ancestor lived some 105 million years ago, it is likely that the same glial scaling rules seen today applied at that time—and have been conserved since then.

with decreasing neuronal density, that is, with increasing average neuronal cell size.

And indeed it did—and not only in the cerebral cortex: again, we found a single, universal relationship between decreasing neuronal density and increasing glia/neuron ratios not only across mammalian orders and species, but even across different brain structures—and, once more, humans fit right in. As shown in figure 9.7, the glia/neuron ratio in a brain structure—*any* brain structure—was a predictable, universal function of neuronal density in that structure: the lower the neuronal density, the larger the average size of neurons in the tissue, and the higher the glia/neuron ratio in that tissue.

I can't stress enough how extraordinary it is to find something that hardly changes in evolution, when the very word "evolution" means change over biological time. If something about animals doesn't change as they evolve, either there is a fundamental physical constraint—such as in the universal relationship between surface area and volume when the shape

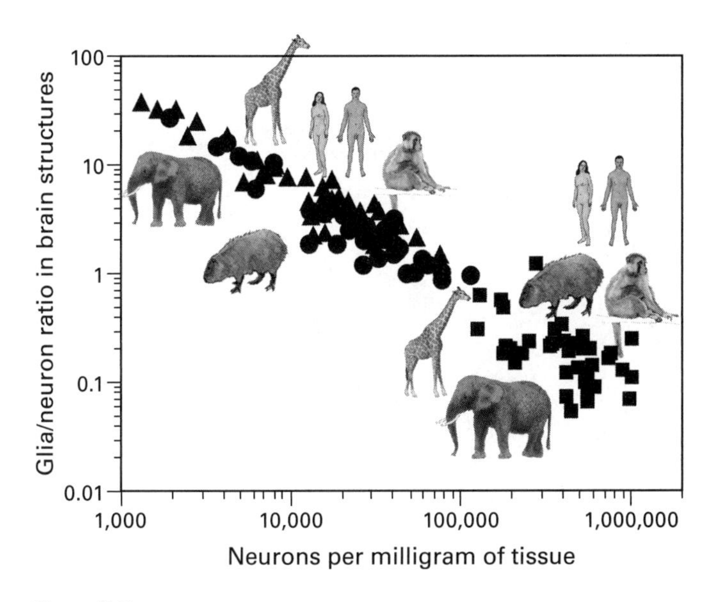

Figure 9.7
Glial/neuron ratio varies as a single, universal function of neuronal density (in neurons per milligram of tissue) in all brain structures, species, and mammalian orders in our data set. For each species, there are three data points in the graph (cerebral cortex, circles; cerebellum, squares; rest of brain, triangles).

of a body doesn't change, or between the cross-sectional area of legs and the mass of the body they support in quadrupeds, both noticed by Galileo—or else there is a biological constraint, such as the fact that all life is based on electrochemical gradients and shares the same genetic code (that is, the same correspondence between DNA bases and amino acids).

In the case of the cellular composition of the brain, we had found not one, but *two* universal features, properties that have remained unchanged over evolutionary time: the number of glial cells per unit of tissue mass (glial cell density) and the relationship between the glial/neuron ratio and the average size of neurons, no matter how greatly neurons varied in number and mass (and we calculated that they varied by at least 200 times across species and brain structures). Given what we knew by then about the formation of brain tissue in development, we argued that both features could be predicted by a scenario in which glial cells were added in self-regulating numbers and only slightly variable cell size.

The model becomes intuitive when we consider that, in the development of each brain, glial cells are only added to the brain in large numbers after neurons have already established the parenchyma, the brain tissue itself.[29] Thus, neurons come first—and neurons, as judged from the large variation in neuronal densities, came in all different sizes, both within a single brain and across species, as illustrated in figure 9.8, where the same

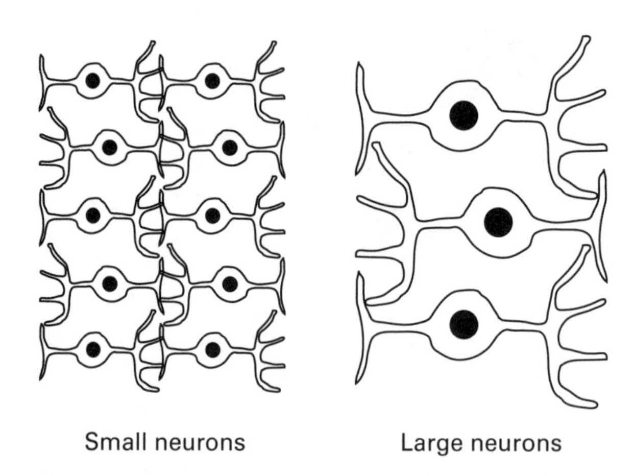

Figure 9.8
Same volume of brain tissue can be composed of a large number of small neurons (*left*) or a small number of large neurons (*right*).

volume of tissue could be composed by a very large number of tiny neurons, as on the left, by a very small number of very large neurons, as on the right, or by any combination in between. The first scenario is, for example, found in the cerebellum, with its immense population of very small granule cell neurons; the second is found in the cerebral cortex of large artiodactyls; the third, in the cortex of rodents, for instance.

Once established, this neuronal volume is then invaded by precursor cells that divide and give rise to the numbers of glial cells that will fill up the tissue. What our data indicated was that the average size of these glial cells did not vary a whole lot across structures;[30] their numbers were simply proportional to the volume (or mass) of the tissue—like the elastic bean bag in chapter 4, which increased in volume proportionally as it gained more Styrofoam beads. The key extra fact here is that the division of glial cell precursors is *self-regulating*: once precursors and daughter cells touch, further division of the progenitor cells is inhibited. This is a self-organizing way of determining how many glial cells fit in a given volume of brain tissue that works exactly like filling an elevator to capacity.

Sure, how many people fit into an elevator can be predetermined by a sign establishing a commandment, say, "no more than 8 people shall ride in this elevator at a time," and even enforced by a supervisor. But glial cells and their precursors don't need signs or supervision, and neither do people. No one even needs to "know" anything: commandment or not, once enough passengers have come into the elevator, they touch shoulder to shoulder—and this is the signal that there is no more space available for another rider. The number of riders in an elevator is, therefore, a self-regulated feature of elevators that depends on the average size of the passengers (and the size of the elevator). As with passengers, so with glial cells and their precursors: once they "touch shoulder-to-shoulder," which indicates that they have entirely filled the volume that was once only neuronal, they automatically stop dividing. No outside supervision required.

And how many glial cells fit in that once neuronal-only brain tissue? If glial cells are all roughly of the same average size, then their final number will always be the same in a given volume of brain tissue—as with riders of the same average size in an elevator of the same size filled to capacity. Of course, there is some small variation, depending on the actual size of the individual glial cells entering the brain tissue or of the individual riders entering the elevator. But, provided there is no systematic variation across brain tissue and glia cells or across elevators and riders, the number of glial cells or riders in this scenario will always be proportional to the volume of the brain tissue or the size of the elevator.*

And so the universal relationship between brain structure mass and number of glial cells can be explained by a self-regulating mechanism that determines how many glial cells fit in a previously neuronal-only parenchyma, provided that the average size of glial cells varies little with tissue mass (and this, in turn, is a function of numbers of neurons and their average size). This universal relationship is therefore the result of simple mechanisms acting on what must be a biological constraint: the very small variation of average glial cell sizes. So that part is accounted for. But what about the glia/neuron ratio?

It turns out that this ratio is also explained by the same model. As shown in figure 9.9, if the final number of glial cells in a given brain tissue is determined simply by the volume of the tissue, which in turn depends on the product of the number of neurons and the average size of the neurons in the tissue, then a given tissue volume made up of a large number of small neurons (on the left) will be filled by the same number of glial cells as the same tissue volume made up of a small number of large neurons (on the right). In the first instance, the glia/neuron ratio necessarily turns out to be low, whereas in the second, the ratio is high. The key factor determining the glia/neuron ratio in a brain tissue is how large the average neuron is. How much energy the average neuron in the tissue costs may or may not be relevant—and it most likely is *not* relevant, as we will see next.

How Much Does a Neuron Cost?

At this point, we knew that the glia/neuron ratio scaled with the average size of the neurons in the tissue: larger neurons were indeed accompanied by larger proportions of glial cells for every neuron, just as Hawkins and Olszewski had supposed in 1957. Did that, however, have anything to do with the increased metabolic cost of larger neurons that required more glial cells per neuron to support them energetically, as they also supposed?

*Glial cells are of different types and sizes (astrocytes, oligodendrocytes, microglia), but since in principle this self-regulation should apply to all subtypes, the end result is the same.

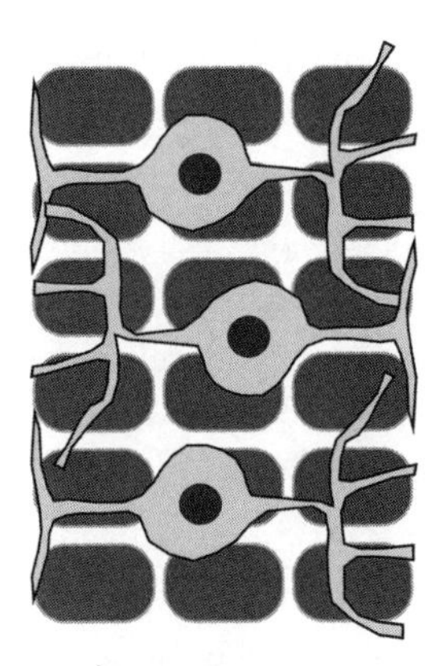

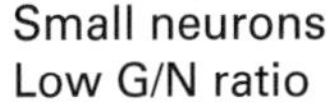

Large neurons
High G/N ratio

Figure 9.9
When the average size of glial cells (the rounded rectangles in the schemes) is nearly invariant, the number of glial cells per given volume of brain tissue is constant. However, because the same volume of tissue can be made of a large number of small neurons (*left*), a small number of large neurons (*right*), or any combination in between, the glia/neuron ratio in each tissue depends simply on the average size of the neurons in the tissue: the larger the average size of neurons, the higher the glia/neuron ratio in the tissue.

It seemed to be a reasonable supposition. Because neurons are, by definition, excitable cells, capable of depolarizing and repolarizing their membrane potential, larger neurons are expected to cost more energy at the very least because of the larger surface area of excitable cell membrane that needs to be repolarized following the depolarization that characterizes neuronal activity. Although depolarization is cost free, akin to opening the gates of a dam, repolarizing requires energy—like a pump requires energy to push water uphill, back into the dam.

Additionally, larger neurons are presumed to have more synapses, and excitatory synapses that use glutamate as a neurotransmitter incur a high energy cost for recycling glutamate and refilling synaptic vesicles.[31] Indeed, a seminal study by David Attwell and Simon Laughlin[32] estimated that nearly 80 percent of a neuron's energy budget goes toward glutamate-related neurotransmission and that 13 percent is used simply to maintain the resting potential of its cell membrane. Although the massive cost of excitatory synaptic transmission could, in principle, be kept down by regulating neuronal activity and numbers of synapses, there is no getting around the requirement for more energy to repolarize the cell membrane of a larger cell once it is activated, that is, depolarized. A large neuron must use more energy to keep its larger membrane surface area polarized—because the cost of *not* repolarizing, due to the large amount of calcium released from internal stores when a cell is depolarized, is, well, death.

Do larger neurons actually cost more energy, though? Testing that hypothesis required determining both how much energy different neurons cost and their average size. As much as I wanted to know the answer, I was not about to embark on a whole new enterprise to measure how much energy different neurons in different brains cost.

Luckily, the literature came to the rescue: a number of researchers had, for unrelated reasons, been able to place awake, unanesthetized adult animals of six different rodent and primate species in positron-emission tomography (PET) scanners to measure the rate at which their brains consumed glucose and oxygen. Even better, the data had been tidily organized by Polish physicist Jan Karbowski, who examined how brain metabolism scales with brain mass (but not with number of neurons). Karbowski[33] confirmed that the mass-specific metabolism of the brain (its glucose use per gram of brain tissue per minute) declines with increasing brain mass across mammalian species, as a power function of brain mass with the exponent –0.15. With such scaling, a ten times larger brain costs only 70 percent as much energy per gram of brain tissue as the smaller brain, which could be due to a smaller number of larger neurons in the tissue but might also indicate decreased average firing rates in the larger brain, in Karbowski's view. Such a larger brain would still cost more energy, overall, a total of 7 times more energy than the 10 times smaller brain: the product of 10 times more tissue using only 70 percent as much energy per gram of tissue. Karbowski thus found that larger brains cost a total amount of energy that grows proportionally with brain mass raised to the power of +0.85 across the species available. This means that the total metabolic cost of the brain increases at a faster rate than the metabolic cost of the body as a whole, which Max Kleiber had shown that increases with body mass raised to the smaller power of +0.75. It seemed that to grow a larger brain was even more expensive than to grow a larger body, which might help explain why brains do scale in mass more slowly across species than the body, as we saw in chapter 8.

The only problem was that Karbowski's study did what we were just learning was no longer justified: it lumped primate and rodent species together as if they shared the same neuronal scaling rules, that is, the same relationships between number of neurons, brain size, and neuronal density—which we were finding was not the case. Moreover, we now knew how many neurons composed each of the brains in Karbowski's analysis, for which data on glucose and oxygen use were also available, and we knew the average neuronal densities in the different brain structures, from which we could infer the average neuronal cell size. I was thus, for the first time, in the position to find out how average metabolic cost per neuron scaled with average neuronal size: did it really increase, as predicted by Hawkins and Olszewski?

I used the same data that Karbowski had compiled, on the energy cost of the awake, unanesthetized brain of three rodents (mouse, rat, and squirrel) and three primates (macaque, baboon, and human). By 2010, knowing the numbers of neurons that composed the brains of each of these species, I could do very simple math and calculate the average metabolic cost per neuron in each of the six species, both for the cerebral cortex and cerebellum separately and for the brain as a whole.

What I found was surprising: the estimated average glucose and oxygen use per neuron within each structure was remarkably constant across the six species. Although the number of neurons varied by a factor of 1,200 across the different brains, and their average cell size varied by a factor of about 3, the estimated average glucose use per neuron varied by a factor of only 1.4, from 4.93×10^{-9} micromole of glucose per neuron per minute in the macaque to 7.05×10^{-9} micromole in the squirrel.[34] The average metabolic cost of the human brain fell in between, at 5.44×10^{-9} micromole of glucose per neuron per minute. This translates into 3.3 billion molecules of glucose consumed on average per human neuron per minute,* which seems like a startling number. Put into perspective, however, it is actually not much. At the rate of 3.3 billion molecules of glucose per neuron per minute, 1 gram of glucose—one-fourth of a

*The math here is straightforward: because, by definition, there are 6.022×10^{23} (Avogadro's number of) molecules in one mole of any substance, and because 5.44×10^{-9} micromole = 5.44×10^{-15} mole of glucose, these many moles thus contains $5.44 \times 10^{-15} \times 6.022 \times 10^{23} = 32.75968 \times 10^{8}$ or 3,275,968,000 molecules of glucose.

teaspoon—contains enough glucose molecules to feed all 86 billion neurons in the human brain for 12 minutes, and 5 grams or 1¼ teaspoons of glucose, enough molecules to power the entire human brain for one full hour!

Not only was there not much variation in the average energy cost per neuron across species—the average human neuron cost only slightly more energy than a macaque neuron, though still slightly less than a squirrel neuron—but there was also no significant correlation between neuronal density (a proxy for the inverse of neuronal size) and the average energy cost per neuron in the brain structures (figure 9.10). Larger neurons did *not* cost more energy.

That was a huge blow to the theory that larger neurons were accompanied by more glial cells because they required more energy: the higher glia/neuron ratio that accompanied larger neurons in any brain structure,

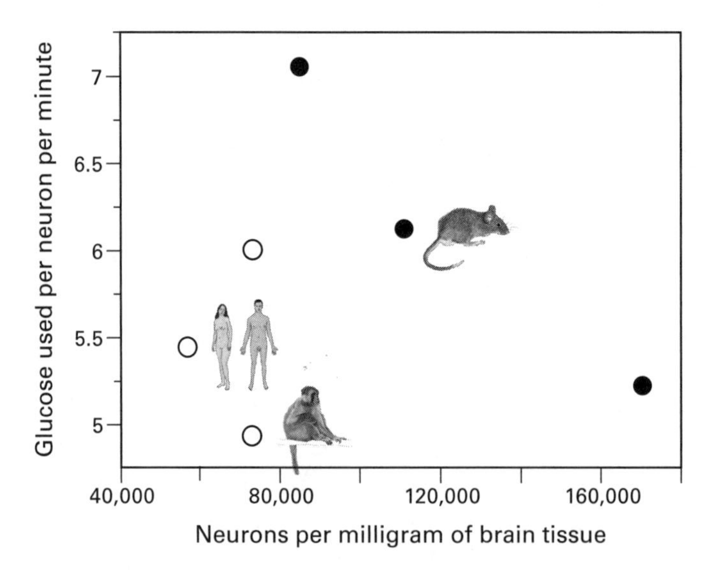

Figure 9.10
Average energy cost per neuron in the brain (in micromoles of glucose per neuron per minute) varies little across mouse, rat, squirrel (filled circles), macaque, baboon, and human (open circles) and, most important, with no obvious correlation with average neuronal density in the brain: larger neurons (at lower neuronal densities) do not use more glucose per minute than smaller ones (at higher neuronal densities).

in any species, could not be related to an increased need for metabolic support.

Most important, because the average energy cost per neuron varied little across species, the total energy cost of each brain turned out to scale as a simple, almost perfectly linear function of the total number of neurons in the brain across primates and rodents alike, as seen in figure 9.11—despite the different scaling rules that related number of neurons to brain mass across those groups. The more neurons in a brain, the more energy that brain costs, as a simple proportionality.

This simple proportionality is illustrated by the fact that a mouse brain, with 71 million neurons, costs 0.37 micromole of glucose per minute; a macaque brain, with nearly 100 times as many neurons (6.4 billion), costs nearly 100 times as much glucose per minute (31.4 micromoles); and the human brain, with just over 1,200 times as many neurons as a mouse brain, costs just over 1,200 times as much glucose as a mouse brain: 497.9 micromoles per minute. Over a full day, the 71 million neurons in a mouse brain

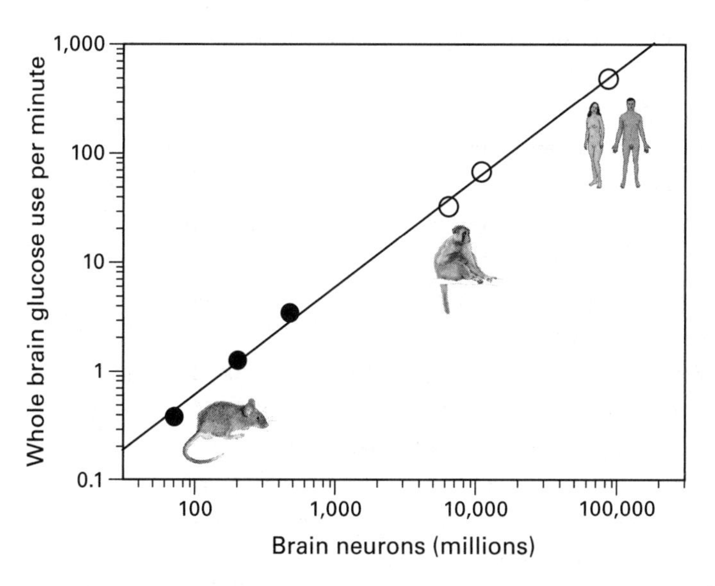

Figure 9.11

Total energy cost of the brain (in micromoles of glucose per minute) scales as a single linear function across mouse, rat, squirrel (filled circles), macaque, baboon, and human (open circles). That is, the more neurons in a brain, the more energy it costs, in a simple proportion.

cost 0.11 grams of glucose, or just 0.4 kilocalorie per day—something that an urban mouse can obtain with just one bite of a cookie (if it can find one). A macaque brain, with 6.4 billion neurons, costs 9.6 grams of glucose per day, or a still relatively meager 38 kilocalories. And, by simple proportionality, a human brain, with its 86 billion neurons on average, requires 129 grams of glucose to work in a day, or 516 kilocalories. A single cupful of sugar provides that many calories.

The average energy consumption per neuron per minute in the cerebral cortex, at 15.0×10^{-9} micromole of glucose, is almost 20 times that in the cerebellum, at 0.9×10^{-9} micromole of glucose, and this proportion between the two structures is fairly constant across the six species examined. The difference could be due to sheer average neuronal size, which is much larger in the cerebral cortex than in the cerebellum in each of the six species. At first glance, this seems at odds with the finding that, *across species*, larger neurons do not cost more energy. Still, a much larger neuronal cell type, such as a cortical neuron, can be expected to have many more synapses than a small neuronal cell type such as the tiny granular cell neurons in the cerebellum. And since the energy cost of a neuron is expected to depend heavily on its number of glutamatergic, excitatory synapses, large neuronal types with many synapses should as a rule cost more energy than small neuronal types with few synapses. What our findings indicate, however, is that *within* each of these neuronal types, cerebellar or cortical, variations in cell size across species are accompanied by neither an increased nor a decreased energy cost: neurons of a same type still cost the same across species, regardless of their size, with a fixed average energy budget per neuron across species.

A little bit of background knowledge on the scaling of energy cost of other cell types across species helps put the meaning of a fixed energy budget for neurons in perspective. Although it makes intuitive sense that a larger neuron should cost more energy, biologically, that is actually quite a stretch. In other body organs, such as the liver, the intrinsic metabolic activity of the cells actually *decreases* with increasing body size.[35] By the same logic, then, larger neurons, if anything, should also use *less* energy, not just as much as smaller ones, and certainly not more. The relatively constant values of energy use per neuron across mammalian species having lower neuronal densities (and hence larger neurons) suggests that the energy budget per neuron, contrary to the energy budget for other cell types, has been stretched close to its limit, imposing a constraint on neuronal activity. This scenario reconciles the low metabolic cost per gram of mammalian brain tissue with the findings of comparative genetic analyses: genes related to cell metabolism are among those that exhibit the larger changes in human evolution,[36] which suggests that maintaining a constant metabolism in brain neurons is rooted in evolutionary genetic changes.

Additionally, there is evidence that the energy budget of individual neurons is not only limited across mammalian species, regardless of neuronal cell size, but also limited over time within a single brain, given that it does not accommodate large variations related to neuronal activity. Although increases in neuronal firing frequency incur directly proportional increases in energy use by the human brain,[37] these are very small. "Activation" of human brain regions upon somatosensory stimulation, which appears as bright red blotches on magnetic resonance imaging (MRI) images, suggests strongly increased use of energy in the recruited brain regions, but such stimulation causes only a 5 percent increase in the metabolic rate of the awake human somatosensory cortex,[38] and visual stimulation causes, at most, only a 8–12 percent increase in the metabolic rate of the awake human visual cortex.[39] Typically, "brain activation" indicated by those red blotches on MRI images amounts to no more than 2–5 percent increases in local energy consumption. It is amazing how much we are able to do with such a slight variation in energy use by brain neurons.

The neuronal energy budget also appears to be both limiting and critical for the maintenance of the conscious state of the brain. Because provision of energy to brain neurons is directly dependent on blood flow, even decreases of 1 percent in overall blood flow compromise brain function, leading to dimmed or blackened vision and to unconsciousness. When we stand up too fast for the reflex that temporarily raises our blood pressure to maintain a constant blood flow to the brain, the large numbers of neurons in the visual cortex are the first to suffer the sudden lack of oxygen to break down glucose—and vision dims or turns black. Similarly, the approximately 45 percent reduction in glucose or oxygen consumption with anesthesia-induced loss of consciousness is compatible with the idea that maintaining consciousness is heavily dependent on energy availability.[40] The energy budget available for neurons seems to be just barely enough for the maintenance of healthy brain activity compatible with waking, awareness, and consciousness: neurons are constantly pushing the limit. It is no wonder that the brain is particularly sensitive to restrictions in blood supply that compromise the availability of glucose and oxygen. For the same reason, chronic impairments of neuronal metabolism would be expected to compromise brain function and contribute to brain pathology, which may be the case in epilepsy (due to runaway excitatory activity or to neuronal metabolic disturbances), mitochondrial disorders (where disrupted energy transfer strongly affects the brain), Alzheimer's disease (where neurons become insensitive to insulin and are disrupted in their ability to take up glucose from the blood), and even normal aging (as neuronal metabolism becomes increasingly compromised[41]). Keeping neurons at the limit of their energy budget has its consequences—and helping neurons stay at that limit in the face of illnesses and aging is a novel and promising approach to treating brain disease.

Living on the edge, at the limit of the energy budget, should also have consequences for healthy neurons. If larger neurons incur an obligatory larger energy cost simply because of the larger membrane surface area to keep polarized, but still have to do within a limited energy budget that is relatively invariant across species, then larger neurons must necessarily cut other costs. For example, mechanisms must be in place that decrease firing rate with increasing neuronal size and thus avoid excessive synaptic activity and the frequency with which the cell membrane must be repolarized. Indeed, larger neurons with larger numbers of synapses have been found to have sparser connectivity and reduced unitary synapse strength, such that firing rate is preserved across clusters of different sizes, but reduced in individual neurons cultured in the lab.[42] Sparse coding, where only a small proportion of neurons fire at high frequencies at any moment in time,[43] may also be a direct consequence of this limiting, size-invariant, fixed energy budget per neuron.

It is interesting to think that a series of fundamental neuronal properties may be direct consequences of constraints on neuronal activity imposed by a limited energy budget. One such property, "synaptic homeostasis," which describes the adjustment in sensitivity of individual synapses in a neuron over time depending on their level of activity, avoids runaway increases in excitatory synaptic activity and therefore in energy cost.[44] Another, synaptic plasticity, the process of removal of unused or nonfunctional synapses as other new ones are added or strengthened, might be a mandatory mechanism that keeps the total number of excitatory synapses and their energy cost in check. New synapses are added to the brain throughout life—but necessarily at the cost of giving up synapses elsewhere. As a result, learning continues to be possible throughout the lifespan.

So Why Is the Human Brain So Expensive?

Knowing the number of neurons in different brains provided, for the first time, a simple, straightforward answer to the question of why the human brain costs so much energy: because it has so many neurons. At an average cost of 5.79×10^{-9} micromole of glucose per neuron per minute in rodents and primates alike, the human brain costs exactly as much as it would be expected to cost. Our brain turns out to be neither special nor extraordinary in its energy requirement.

The average energy cost per neuron that I found to apply equally across rodents and primates allows the overall metabolic cost of other mammalian brains to be inferred from their numbers of neurons, using the

Table 9.1
Energy cost of mammal brains compared across species

Number of neurons	Total glucose use per day (g/day)	Total caloric cost per day (kCal/day)
1 million	**0.0015**	**0.006**
10 million	**0.015**	**0.060**
Smoky shrew, 36 million	0.05	0.2
Mouse, 71 million	0.11	0.4
100 million	**0.15**	**0.6**
Rat, 200 million	0.30	1.2
Marmoset, 636 million	1.0	3.8
Agouti, 795 million	1.2	4.8
1 billion	**1.5**	**6.0**
Owl monkey, 1.5 billion	2.2	9.0
Capybara, 1.5 billion	2.2	9.0
Macaque, 6.4 billion	9.6	38
10 billion	**15**	**60**
Baboon, 11 billion	16	66
Orangutan, 30 billion	45.02	180
Human, 86 billion	129	516
100 billion	**150**	**600**

estimated average cost of 5.79×10^{-9} micromole of glucose per minute per neuron—as long as they obey the general rule of four neurons in the cerebellum to every neuron in the cerebral cortex. This cost is equivalent to 5.79 micromole of glucose per minute per billion neurons per day, or 6 kilocalories per billion neurons per day—a handy round number to remember and use to infer the energy cost of different brains, as shown in the table above.

For species such as the African elephant, whose cerebellum has an extraordinarily large proportion of brain neurons, however, the energy cost of the cerebral cortex and cerebellum must be calculated separately using the average glucose consumption found for each structure. Although the cerebral cortex accounts for roughly 50 percent of the energy cost of the human brain (presumably as in almost all other mammals), the extraordinarily large number of neurons in the elephant cerebellum is predicted to cost almost three times more than the neurons in the elephant cerebral cortex. With such a high metabolic cost, all those neurons in the elephant cerebellum must really serve a very important purpose.

So the human brain is not special in the absolute energy cost of its neurons (it costs just as much as expected for its number of neurons), nor in the relative cost of its cerebral cortex (which holds a similar proportion of all brain neurons as in other species, elephant excluded). Why, then, is the human brain's *relative* energy cost so high, reaching 25 percent of the energy used by the entire body, whereas in other species it amounts to 10 percent at most?

The answer here is, once more, that humans are primates—and, as such, have a much larger number of brain neurons for their body mass than other, nonprimate species. Figure 9.12 shows the relationship between brain neurons and body mass translated into the predicted energy cost of the brain (using the estimated average cost of 6 kilocalories per billion neurons per day) and the predicted energy cost of the body (from Kleiber's law).* For a similar body mass, which supposedly costs just as

*Daily energy cost of the body in kilocalories = $70 \times (\text{body mass in kilograms})^{+0.75}$. Thus, for the 70-kilogram human body, the daily energy cost would be $70 \times (70)^{+0.75}$ $= 70 \times 24.2 = 1{,}694$ kilocalories $\approx$ 1,700 kilocalories. See Kleiber, 1932.

much energy for a rodent as for a primate, the brain of a primate is more expensive compared to a nonprimate brain simply because it has more neurons.

And within primates, the human brain is the one with the largest absolute metabolic cost simply because it has the most neurons. We found that the absolute energy cost of the brain grew linearly with the energy cost of the body in primates (as a power function with the exponent 1.0) but sublinearly with the energy cost of the body in nonprimates (as a power function with the exponent +0.6; figure 9.12). As a consequence, the *relative* metabolic cost of the brain, though it varied quite a bit, did not vary systematically with body mass across primate species, but it became progressively lower as body mass increased across nonprimate species (figure 9.13). And, as it turns out, the human brain does *not* have the highest *relative* energy cost: that honor goes to the tiny squirrel monkey's

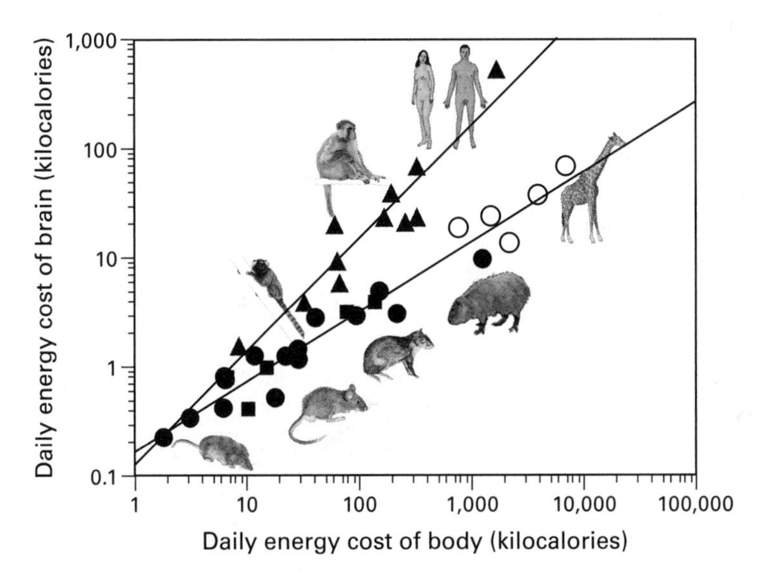

Figure 9.12
Total daily energy cost of the brain (in kilocalories), predicted to vary across species as a simple, linear function of the total number of neurons in the brain (at an average cost of 6 kilocalories per billion neurons per day), scales as different functions of the daily energy cost of the body (also in kilocalories) across primates (triangles; with the exponent +1.0, that is, linearly) and nonprimates (squares, open and filled circles; with the exponent +0.6).

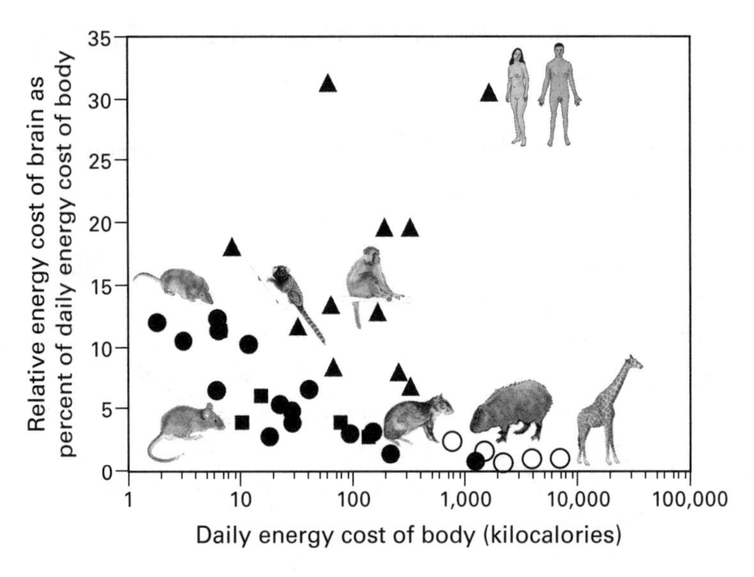

Figure 9.13
Relative daily energy cost of the brain, expressed as a percentage of the daily energy cost of the body (in kilocalories), becomes lower with increasing body mass across nonprimates (squares, open and filled circles), but varies nonsystematically across primates (triangles)—and is predicted to be even higher in the squirrel monkey than in the human.

brain, which, with a particularly high number of neurons in the cerebral cortex, consumes an estimated 31.2 percent of the daily energy used by the monkey's entire body, as compared to the human brain's estimated relative energy cost of 30 percent, given a 70-kilogram (155-pound) body that consumes 1,700 kilocalories per day.

It becomes clear that brain metabolism is not related to whole body metabolism in any single determining way: any apparent relationship might be coincidental and dependent on the rate with which brain size scales as a function of its number of neurons, which we have shown to vary across mammalian orders. Primates, whose brain evolved with economical neuronal scaling rules, have a large number of brain neurons for a given body size, and their brains are accordingly expected to have a higher relative metabolic cost than the brains of other mammals, such as rodents, which have a smaller number of neurons for the same body or brain size.

Once more, it turns out that there is a simple explanation for something that appeared to be extraordinary about the human brain. It is relatively expensive compared to the body simply because it is a primate brain, and, as such, it holds a large number of neurons for its body compared to nonprimates. But the human brain still costs just as much energy as expected for its number of neurons, and just as much energy compared to the body as expected for a non–great ape primate of its body mass. Though remarkable in its energy cost because of its remarkable number of neurons, it is, once more, not special.

10 Brains or Brawn: You Can't Have Both

How expensive the human brain is in terms of energy cost depends on how one looks at it. On the one hand, it costs around 500 kilocalories a day, which seems enormous compared to the total 2,000 or so kilocalories that the entire human body consumes per day—even when it turns out that this cost is just the expected one for the remarkable number of neurons packed in the human brain, which in turn is just the expected number for a primate brain of its size. On the other hand, 500 kilocalories per day translates roughly into 24 watts of power: all that we mentally achieve is done while using just over one-half the energy it takes to power a 40-watt lightbulb, and just over one-third the 60 watts it takes to power a laptop computer. The rate at which the human brain consumes energy, that is, its power, is steady at about 24 watts. In comparison, our muscles have variable power and can work at a rate up to three times as large as the brain—75 watts—and even more in short bursts, or in athletes. Heavy lifting requires more energy, but heavy thinking, surprisingly, consumes no more energy as a whole than, say, mental idling. To be sure, some parts of the brain become slightly more active and others, less, but the redistribution of blood flow from the less active to the more active parts explains how its overall energy cost is the same whether the brain is idling or focused. Seen this way, the human brain is a remarkably efficient machine, all the more so because it can be powered for one full hour by 1¼ teaspoons (about 5 grams) of glucose.

And yet when we consider the energy all the neurons in the brain consume over the 24 hours of the day, the remarkable number of neurons in the human brain, and in particular in its cerebral cortex, is so expensive that other primates simply can't afford anywhere near as many of them. With its 86 billion neurons, on average, the human brain requires 129 grams of glucose or 516 kilocalories to run for a day. A cupful of sugar has that many kilocalories, and consuming that is a feat so simple for humans in our urban, refrigerator- and supermarket-filled world that we face the opposite problem to finding calories: we consume too many of them.

If only it were that simple to find calories in the wild.

Energy In = Energy Out

Animals that we are, we must face, every single day of our lives, the consequences of our most basic predicament: we don't do photosynthesis. For lack of the necessary genes, we don't just absorb carbon from the air around us and fix it as new bodily matter with a little help from sunlight. To survive, we animals have to eat other living organisms, whether animal, vegetable, or fungus, and transform their matter into ours.

Just as our cars run on gasoline and cannot start on an empty tank, so our bodies run on curiously similar organic compounds that are broken down and have the energy of their chemical bonds harvested in our mitochondria, then redistributed inside the cell. The difference is that, though we may worry that the gas tank in our car is nearly empty and the next gas station is miles away, most of us in the urbanized world are blissfully unconcerned about *when* we'll be able to find fuel for our bodies and eat again. We give thought to what exactly we would *like* to eat, and maybe where; but most of us who live in the modern cities of the world take it for granted that food will be available on demand.*

But in the wild, where our ancestors came from, having food to eat is not at all guaranteed, and looking for it can take a lot of time and effort—so much so that the effort has its own name: "foraging." Although "foraging" in the home is as simple as going to the kitchen cupboard and taking a few handfuls of chips from a bag or three cookies from a jar for an "intake" of 150 kilocalories in not much more than a minute or so, obtaining that

*As very aptly put by Douglas Adams in *The Restaurant at the End of the Universe*: "The History of every major Galactic Civilization tends to pass through three distinct and recognizable phases, those of Survival, Inquiry, and Sophistication, otherwise known as the How, Why, and Where phases. For instance, the first phase is characterized by the question 'How can we eat?' the second by the question 'Why do we eat?' and the third by the question 'Where shall we have lunch?'"

many kilocalories in the wild might take at least an hour—and use up a significant amount of energy in the process.

Actually, depending on the size of one's mouth, ingesting those 150 kilocalories in the wild might take much longer than a minute. With a mouse-sized mouth, there is only so much of a 50-kilocalorie cookie that can be eaten in a minute, whereas a human-sized mouth can devour up to three 50-kilocalorie cookies in that same minute.* Although a gorilla might be able to ingest as many as 300 kilocalories per hour spent foraging (the equivalent of six cookies), a small monkey or marmoset doesn't ingest much more than 10 kilocalories in the same hour.[1]

In the real wild world, away from kitchen cupboards and cookie jars, refrigerators and supermarkets, caloric intake is so limited that the more energy needed to feed a body, the more time that will have to be spent foraging.[2] Caloric intake is further dependent on the availability and quality of foods: orangutans, for instance, spend 7–8 hours per day feeding year round, but, during the months of low fruit availability, those hours are still not enough to provide all the kilocalories required, and the animals lose weight.[3] As we saw in chapter 9, having more neurons comes at a proportionately larger energy cost, which, in principle, requires that more time be spent foraging and feeding in order to support the brain alone.

Only in principle, however. More neurons make for larger brains, which usually come in bodies that are even larger. And though larger bodies also cost more energy per day, they also come with larger mouths that could mean an increased food intake per hour, perhaps even larger than required to keep up with the increased energy needs of a larger body and brain. But does the caloric intake scale up fast enough to meet the increasing energy needs of a larger body, or do the needs of a larger body eventually make it prohibitively expensive?

The reason for my sudden interest in food intake, which seemed to be such a long way from the question of how brains are made, was my

*But no more than three cookies per minute—and possibly not even three. Although it may seem a trivial task to gobble down three cookies in 60 seconds, swallowing is limited by the rate saliva is produced, and, try as one might, it is just not possible to swallow without enough saliva. The reason I'm familiar with this seemingly useless bit of trivia is that eating three cookies in under one minute was a challenge in a competition between freshmen and seniors every new term in my college, one that nobody ever met. Hey, at least it wasn't a drinking game.

discovery of how energy-expensive neurons are—and I had a suspicion that the high energy cost of primate brains in particular, packed with neurons as they are, might explain why gorillas don't have brains nearly as large as we might expect them to have with their enormous body size.

Here, in a nutshell, is what I suspected then, and can now support with evidence. The reason why humans have long been considered special—outliers in comparison to other animals and primates in particular—is that great apes have traditionally been thrown in the mix, when *they* have in fact always been the outliers, not us. Compared to all other primate species for which we have data on body mass and number of brain neurons, both great ape and human brains are generic, scaled-up primate brains in their numbers of neurons and distribution across brain structures. But when it comes to body size, here is where great apes stand out, not us, in comparison to other primates: whereas humans, in their relationship between *body* mass and number of brain neurons, stand where generic primates would be expected to stand, gorillas and orangutans have brains that are much too small for their bodies.

The issue, then, became finding an explanation for what made great apes diverge away from the pattern that we still share with other primates. Now that we knew that the energy cost of primate brains was particularly high for their volume compared to nonprimate brains simply because of the larger numbers of neurons they packed into the same volume of brain tissue, I had a clear suspicion why great apes were the outliers, without larger brains to match their larger bodies: they simply couldn't afford the energy cost of both.

Caloric Intake Is Limiting

Whether an animal is energetically viable depends on having a caloric intake that at least matches the energy needs of its body and brain. What I had in mind to test my idea that a limiting number of available kilocalories forced a trade-off between body mass and brain mass in the largest apes was this. In one pan of a mathematical balancing scale, I would put the number of calories that different animal species—primates, for starters—could obtain from their diet in a single day, and in the other pan, the number of calories required to sustain different combinations of brain mass and number of neurons. I wanted to find out if there was a limit to how many neurons and how large a body a primate could afford, and where that limit was—because if it was a trillion neurons and a body of several tons, well, then it was not a physiologically meaningful limit after all.

Bodies cost a certain total amount of energy per day that can be measured or at least estimated in kilocalories from Kleiber's law, as $70 \times (\text{body mass in kilograms})^{0.75}$, as noted in chapter 9. Since we knew the body mass of a number of primate species, we could easily estimate the total energy cost of their bodies. As to calculating how much energy their brains cost, well, we had the overall average energy cost of 6 kilocalories per day per billion neurons that applied to the brains of the macaque, baboon, and human,[4] so we could assume that the same energy cost applied to the brains of all other primates as well.

What we lacked were estimates of the number of kilocalories that primates obtain per hour when foraging in the wild. But those data were out there, collected by primatologists who had spent many hours observing other primates foraging and feeding. All we had to do was mine and make sense out of their data.

The required data mining was accomplished by an extraordinary undergraduate student in my lab, Karina Fonseca-Azevedo, who had worked with us before on the variation of the cellular composition of the brain across individual mice and had since left the lab. But, upon attending a talk where I suggested that the argument of humans being outliers compared to great apes should be turned upside down—*they* were the outliers, not us—Karina rejoined the lab and asked whether she could work specifically on that question.

And work she did, starting with finding and compiling the data on the average number of hours that different primate species spend foraging and feeding in the wild. Our first question was whether and how the average caloric intake scaled with body mass: were larger primates able to obtain more kilocalories per hour of feeding, as we expected intuitively? And did their capacity to take in calories from food scale *faster* than the total caloric needs of the body, in which case there would be no energetic limitation to becoming larger, or *slower*, in which case primates would eventually hit a wall somewhere that limited their maximal size?

We reasoned that the number of hours that different primates spent foraging and feeding in the wild must be, on average, just enough to take in what they needed to meet their total daily caloric requirements. If they took in any more, adult primates would keep getting fatter and fatter, which is not the case; indeed, wild primates are usually fairly lean. If they took in any less, they would lose weight and be chronically famished, which is also not the case for healthy primates in the wild. With that reasoning, we could assume that the kilocalories obtained by so many hours of foraging and feeding were, on average, just enough to meet their total energy requirements. We could then divide the daily total estimated energy cost of each species by the number of hours spent foraging and feeding per day, and so obtain an estimate of the number of kilocalories that each species took in per hour spent foraging and feeding.

Karina found that this amount, the rate of caloric intake, varied from as little as 9 to 10 kilocalories per hour in small primates such as the marmoset and owl monkey to as much as 202 kilocalories per hour in the orangutan and 335 kilocalories per hour in the gorilla. As shown in figure 10.1, there was a positive correlation between body mass and hourly

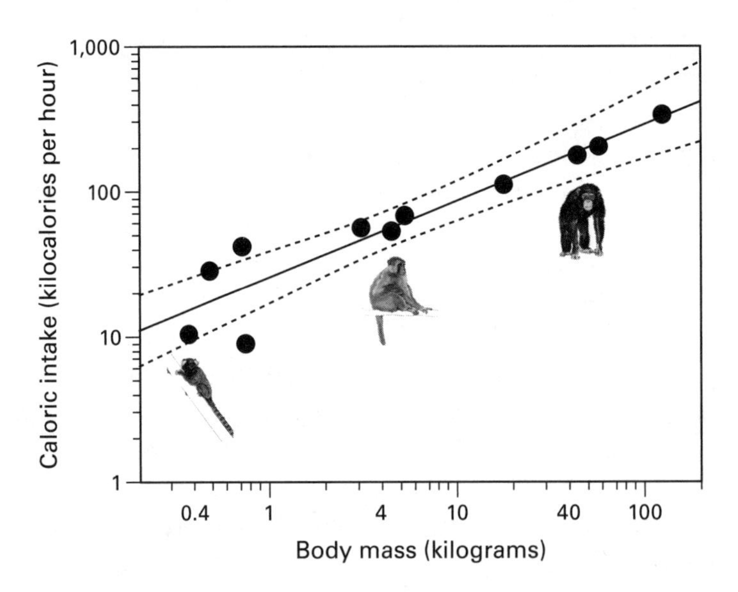

Figure 10.1
Larger primates have an increasing capacity for caloric intake (in kilocalories per hour spent foraging and feeding). However, the hourly caloric intake scales with body mass raised to the power of 0.53 (with dashed lines indicating the 95 percent confidence interval for the scaling function), more slowly than the energy cost of the body, which scales with body mass raised to the power of 0.75.

caloric intake, as expected: larger primates could amass more kilocalories per hour of foraging and feeding. There is a definite advantage to having a larger mouth.

Still, we found that the hourly caloric intake of primates in the wild scaled as a power function of their body mass with the exponent 0.53—smaller than the exponent 0.75 for the power function of how the total metabolic cost of the body scaled with increasing body mass. This meant that the energy cost per hour of larger primate species increases *faster* with body mass than their capacity to obtain kilocalories per hour, which meant, in turn, that somewhere there would be a wall limiting how large the body of a primate species could become.

But that only applied to hourly rates, and, in theory, there was always the possibility of foraging and feeding for longer times, if not enough kilocalories could be amassed per hour. Indeed, Karina found that the total number of hours spent foraging and feeding per day increased with body mass: from an average of under 2 hours per day for the smallest primates to over 7 hours per day for orangutans and almost 8 hours per day for gorillas, as shown in figure 10.2.

In practice, however, days have 24 hours, and the number of hours that a primate can spend foraging and feeding is limited further by the requirement to sleep between 8 and 9 hours per day—yet another way in which we humans resemble our fellow primates. Orangutans seem to be limited to a maximum 8.5 hours of foraging and feeding per day, which is the average number of hours they can consistently devote to these activities when food becomes scarce[5]—and still they lose weight during the dry season, which indicates that they would need to spend even longer hours procuring kilocalories under these circumstances—but they can't. A total of 10 hours of foraging and feeding per day has been documented in gorillas,[6] but that seems to be the result of the occasional, extreme effort to find kilocalories, and not something that could be sustained daily.

Using 8 hours as a practical limit to how many hours a primate could spend foraging and feeding, we put in one pan of our scale the daily caloric intake calculated for primates that foraged and fed for 8 hours per day, depending on their body mass, and in the other pan, the daily energy cost of the primates with the same body mass. The maximal body mass that a primate that foraged and fed daily for 8 hours could afford is given by making one equation equal the other to determine when energy taken in equals

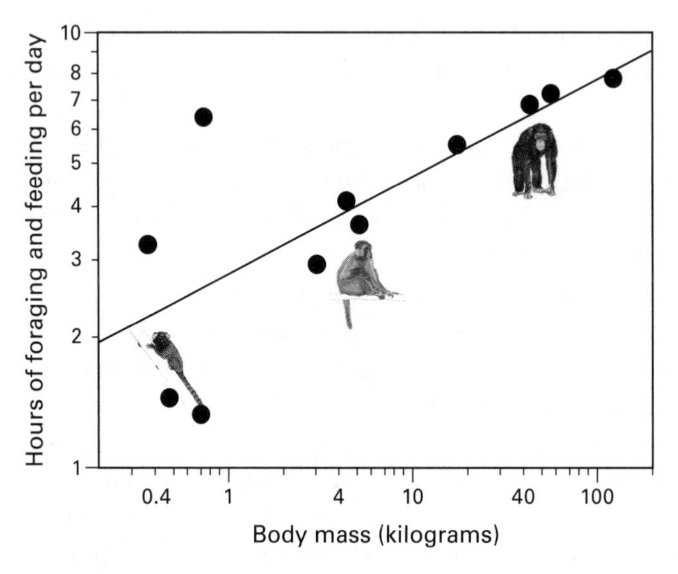

Figure 10.2
Larger primates spend more hours per day foraging and feeding. Although the number of hours per day they spend in these activities scales slowly, with body mass raised to the power of 0.22, the large range of primate body sizes is enough to increase the required number of hours spent amassing kilocalories from less than 2 to almost 8 hours per day.

energy used—and that we found to happen at around 120 kilograms (265 pounds). That this body mass is not enormous but reasonably close to the typical weight of a non–alpha male silverback gorilla in the wild is highly significant: there *is* a limit to body mass imposed by caloric availability, and gorillas live not too far below that limit. Becoming a larger alpha male silverback requires an extra effort to find food—but, then again, one of the perks of being an alpha gorilla is getting food from the dominated peers, so in the end, it pays off to become large.

Brains or Brawn

We hadn't even considered the particular cost of larger numbers of neurons, and we had already run into a meaningful energy limit to how large primate bodies could be. What if we now focused on that particular cost? Since we knew the mass of the brain of each species, we could subtract it from the total mass of the body to estimate the energy cost of a body without the brain (mostly a formality, given how relatively small the brain is compared to the body). Since we knew or could estimate the number of neurons in each of the primate species, we could then add back the number of kilocalories required to run a brain with that number of neurons. Actually, we could estimate the number of kilocalories required to run any primate brain, based simply on the predicted number of neurons for their brain mass, and assuming that the brain used the same 6 kilocalories per billion neurons per day that we had found was the case for the brains of humans, rodents, and two other primate species.[7] We were now in the position to put in one pan of our balancing scale the amount of energy that primates could obtain from their diet (depending on the mass of their body and the number of hours spent foraging and feeding per day) and in the other pan, the total energy cost due to different combinations of "brainless" body mass and number of neurons. Knowing that the rate of caloric intake was already not enough to sustain ever larger primate bodies suggested that adding the energy cost of the neurons to the other pan would only tip the scale faster. The question, at this point, was how soon would primates with a given number of brain neurons run out of hours to compensate for the larger energy requirement by foraging and feeding for longer times each day.

By balancing the equations for caloric intake and caloric use for brain and body, all depending on body mass, number of brain neurons, and number of hours spent foraging and feeding per day, we could establish "viability zones": combinations of body mass, number of brain neurons, and hours dedicated to foraging and feeding that could be sustained on the estimated number of kilocalories amassed under those circumstances, as shown in figure 10.3. The fact that such zones could be delineated at all meant that caloric intake was, indeed, a limiting factor not just to the size of the body, but also to the number of neurons that a primate could afford in its brain, depending on how many hours it foraged and fed per day.

The shape of the viability zones shown in figure 10.3 is all the more important: even for primates that could spend 10 hours per day foraging and feeding, the zones slope downward to the right. This meant that not only was there a limit to how many neurons a primate of a certain body mass could afford (the maximum vertical value of each curve in figure

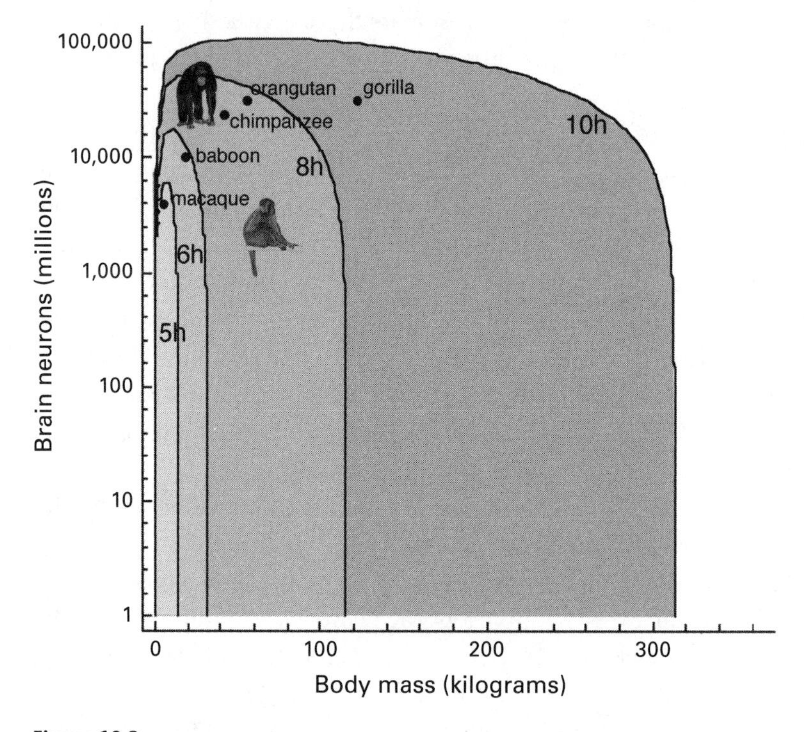

Figure 10.3
Shaded zones for each curve indicate the viable combinations of number of brain neurons and body mass that can be sustained for a given number of hours per day spent foraging and feeding (h). The downward slopes to the right of each curve indicate a trade-off: past a maximum number of brain neurons, body mass can increase only at the expense of the number of neurons; by the same token, the number of neurons at the limit of the curves can increase only at the expense of body mass.

10.3), but that, past a certain point, increasing the mass of the body came at a cost: either foraging and feeding for longer hours (that is, jumping the curves in figure 10.3), or, if that wasn't possible, then giving up numbers of brain neurons.

And this is a steep and physiologically relevant trade-off, one that becomes even more clear when considering the maximal number of neurons that are sustainable on a given number of hours per day spent amassing kilocalories, and the maximum possible body mass to still be able to afford that many neurons, as in the combinations shown below. For example, eating for 8 hours affords a primate a maximum of 53 billion neurons, no more—and in this case, at the cost of limiting body mass to no more than 25 kilograms (55 pounds). Increasing the size of the body while still foraging and feeding for the same 8 hours per day, according to our calculations, was possible, but only at the cost of giving up brain neurons, so that the maximal number of neurons affordable decreases with increasing body mass, as shown below.

53 B neurons	25 kg
45 B neurons	50 kg
30 B neurons	75 kg
12 B neurons	100 kg
not viable	150 kg

Past a certain point, too large a body costs more energy than it can obtain through foraging and feeding for a certain number of hours per day—never mind the extra cost that neurons would have. But too many neurons are also not affordable even for midsized bodies. For a 75-kilogram (165-pound) primate, 30 billion neurons is the limit we calculated. Gorillas and orangutans, with their 33 billion neurons and body mass in the 50–100-kilogram (110–220-pound) zone, are therefore at the limit of number of brain neurons and body mass for a primate that forages and feeds an average of 8 hours per day. The calories they amass could certainly go toward supporting a larger number of neurons—but only if their body mass were smaller. For their actual body mass, they just don't ingest enough calories to afford a larger number of neurons. Other, smaller primates could easily spend more hours per day foraging and feeding, if they needed to, and could afford more neurons over generations, and that must have happened indeed, for there has been a trend toward increased brain and body sizes in the last several dozen million years of primate evolution.

But not anymore once the body had become so large in primate evolution that it required a dangerously large number of kilocalories per day. The orangutan brain is about one-third the size of the human brain, and, therefore, given the linear neuronal scaling rules that apply to primate brains, can be estimated to have roughly one-third as many neurons and to require roughly one-third as many kilocalories to support the brain alone, that is, about 180 kilocalories. During the months of low fruit availability, when total caloric intake by female orangutans is estimated at about 1,800 kilocalories per day (at best, assuming 100 percent caloric efficiency of the foods ingested—which is unreasonable, as we will see soon), their brains are estimated to require at least 10 percent of the total caloric intake, and the remainder is less than sufficient to support the body, so the animals lose weight. Doubling the number of brain neurons in an orangutan brain would have required an additional 180 kilocalories per day, which would take almost one extra hour to amass. Given that orangutans already forage and feed for as long as they can, it becomes clear that any significant increase in the total number of brain neurons would have jeopardized their ability to survive.

The largest great apes, then, can't just spend more time foraging and feeding: they have reached the maximum number of hours they can spend per day in these activities, and thus also the maximum number of kilocalories they can amass per day. A primate can't have both a very large body and a very large number of neurons: it is either brains or brawn—and great apes seem to have "chosen" brawn.

But of course there is no "choosing" in evolution. The workings of natural selection can be recognized only in retrospect, as a given trait proves itself advantageous over generations. The last ancestor shared by humans, chimpanzees, bonobos, gorillas, and orangutans supposedly lived some 16 million years ago. Its body mass is still unknown, but judging from the fossil hominid species that came later, it may have been a midrange primate, possibly chimpanzee sized, though with a brain that already boasted some 30 billion neurons, judging from its cranial size, roughly the same as the modern gorilla or orangutan. From that point onward, the lineages that remained on all fours (and gave rise to modern great apes) seem to have invested any additional kilocalories amassed per day from longer times foraging and feeding into growing larger bodies. For knuckle-walking species that are, for anatomical reasons, not very mobile, becoming as large as they could afford must have been advantageous, earning larger animals higher social status and thus greater access to food, among other privileges.

Certainly, these animals would also have benefited from having even larger number of neurons, assuming that would endow them with greater cognitive abilities to meet their daily challenges. But, as the numbers above show, pushing the limits of those viability curves (figure 10.3) would be a risky business, making starvation and death a real threat. Because an individual brain always uses the same amount of energy, no matter whether the rest of the body is starving, having too many neurons is clearly a liability when a species lives close to the limit of its caloric intake possibilities.

But for our newly bipedal and suddenly highly mobile australopithecine ancestor, who, some 4 million years ago, diverged away from the lineage that would give rise to the modern chimpanzee and bonobo, investing the additional kilocalories it amassed per day in a greater number of brain neurons housed in a leaner, lighter body must have proven a much better investment strategy. How, exactly, the great ape and human lineages came to adopt different investment strategies, privileging brawn over brains or brains over brawn can only be guessed at. What becomes clear, though, once energy requirements and availability are taken into consideration, is that one can't have both.

11 Thank Cooking for Your Neurons

We don't usually think of ourselves as animals, much less as being limited by anything (other than not being able to fly on our own wings). But a primate is what we are, and how we compare with other primates in terms of our energy needs can shed much new light on our evolutionary history. Because if gorillas and orangutans live life on the edge, at the limit of the number of brain neurons and the body mass that primates with their caloric intake can afford, then we humans, well, we shouldn't really be here at all.

That's because the 86 billion neurons in our brains and the some 70 kilograms of our bodies would, by our estimates, require over 9 hours of foraging and feeding every day to be affordable by a generic primate of our body mass, with an average intake of some 200 kilocalories per hour.[1] Foraging and feeding for that long is, of course, something we don't do. Indeed, the typical urban human wouldn't be able to forage and feed anywhere near 9 hours every single day—nor, as it turns out, would our ancestors, either.

Given that we shouldn't be viable, and yet here we obviously are, the pressing question becomes how did our ancestors manage to afford the increasing numbers of neurons that characterized the emergence of modern humans? As figure 11.1 illustrates, one of the hallmark and most remarkable aspects of human evolution is that the brain of *Homo* species increased very much and extraordinarily rapidly—nearly tripled in size during the last 1.5 million years, really—while the brains of our great ape cousins stagnated at the same size they have to this day. To put this into a wider perspective, it took around 50 million years for primate brains to go from microlemur sized to gorilla sized (an increase of 29 billion neurons), but only 1.5 million years to add another 57 billion to *Homo* alone—nearly

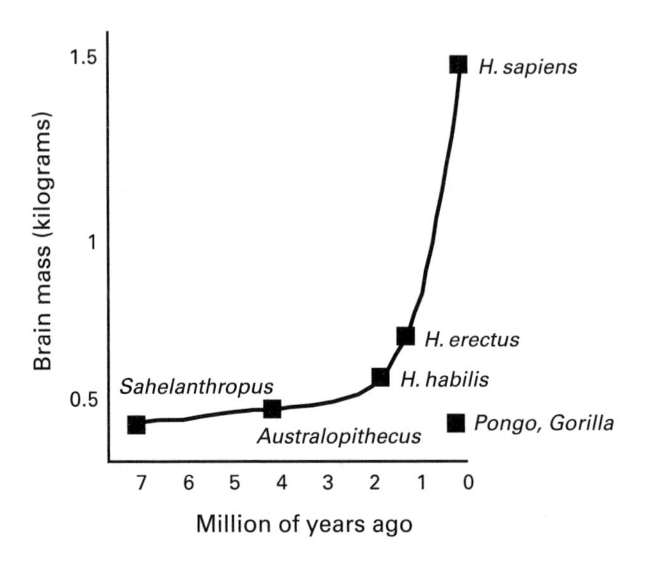

Figure 11.1
Rapid increase of brain mass in the *Homo* lineage in the last 1.5 million years, but not in the lineages that led to modern great apes.

twice more neurons. Compared to that of great apes, our evolutionary history indeed seems extraordinary: an oddity that singles us out.

Just providing the energy for tripling the number of neurons in the brain of *Homo* would, by our and others' accounts,[2] have required a solid 9 to 10 hours of foraging and feeding per day on a typical primate diet such as that of the great apes, which simply isn't something a great ape can do. Gorillas and orangutans lose weight when foraging and feeding for 7.5 hours per day during months of low fruit availability. Our ancestors wouldn't have lasted very long if they had to spend an extra two full hours foraging and feeding per day just for basic sustenance. And if our ancestors couldn't get enough food from their primate diets, we shouldn't be here. Whatever changed that allowed the brains of *Homo*, and of *Homo* alone, to increase so much, so fast, must have been something that made that energetic constraint be a constraint no more.

More Calories in Less Time

There are four ways to work around an energetic constraint to the number of neurons in the brain: (1) decrease the size of the body; (2) decrease the energy cost of the brain; (3) amass more energy by spending even more hours per day foraging and feeding; or (4) somehow increase the energy obtained from the same amount of food—like with a radical change in diet.

Although the human brain was long considered too large for its body (which implied a body too small for its brain), we have seen that, once we exclude great apes from the comparison—and the energetic constraint on their body and brain mass gives us good reason to—humans turn out to have a body and brain that match what applies to most other non–great ape primates, who could spend longer hours per day foraging and feeding if they needed to. So we can rule out way number one: the human body is just the size it should be for its brain mass. Besides, our ancestors were *smaller* than we are, not larger: the human body mass did not decrease in evolution when its brain increased in size.

The human brain also costs just as much energy as it is expected to, given its number of neurons, especially in the cerebral cortex. And foraging and feeding for 9.5 hours per day is not feasible, given that the practical limit for a primate seems to be somewhere around 8 hours per day. So scratch ways numbers two and three, as well.

The remaining way to work around an energetic constraint to the number of neurons in the brain involves dietary changes that would allow more calories to be obtained in the same amount of time, or even less. Some first changes in that direction probably took place as early as 4 million years ago, when our australopithecine ancestors stood upright and became habitual bipeds. As Daniel Lieberman explores in detail in *The Story of the Human Body*,[3] bipedality potentially increases the amount of calories that can be amassed in a day by extending the range of food picking, for it is much easier and costs four times fewer kilocalories to walk on two feet, as humans do, than on all fours, as modern great apes do and as the ancestor from which australopithecines originated must have done. Roaming away from home to find food is the definition of a food gatherer, as opposed to a food picker, which is what great apes remain to this day. Bipedality made food gatherers of our ancestors.

By 2 million years ago, our ancestors already had undergone other modifications compared to the primates who would become great apes: *Homo erectus* had longer legs that reduced the cost of walking, with spring-loaded tendons and muscles that made the cost of endurance running independent of speed.[4] Other traits also benefited endurance running, such as a large gluteus maximus (the muscle that makes human bottoms look round), the nuchal ligament (which keeps the head upright), large semicircular canals in the ears (which help keep balance and gaze steady even while running), and short toes. For a species that lived close to the limit of the metabolically affordable, endurance running facilitated "power scavenging," that is, running considerable distances to a carcass (signaled by vultures in the sky, for instance), chasing away the local carnivores, and running off with what they could carry.

But, even more important, endurance running seems also to have allowed our ancestors, who were fairly small and not that muscular, to add active hunting to their gathering and scavenging. Archaeological evidence indicates that, by 1.9 million years ago, early humans were hunting large animals like wildebeest and kudu—something that, without tipped spears or anything more lethal than a club, and without the muscular power of chimpanzees and gorillas, necessarily involved endurance running, which only bipedal *Homo* could do. Even though unarmed humans couldn't outrun antelopes, in coordinated groups, they could chase them long distances until the animals finally became exhausted and then take them down—especially those humans who, by chance, happened to have a few billion more brain neurons to do the job.

By the time our *Homo* ancestors started to grow much larger brains, they had already become not only gatherers, but also hunters—and hunting, in turn, must have exerted a selective pressure for more brain neurons, requiring still more cooperation, which relied on memory, planning, reasoning, self-control, awareness of the mental state of fellow hunters, communicating through some sort of language: cortical abilities that rely heavily on the associative functions of a prefrontal cortex. Through making more energy available, becoming hunter-gatherers probably put our ancestors on the path toward both benefiting from and being able to afford greater number of neurons in the brain.

But all this happened some 4 to 1.5 million years ago, a period when brain mass increased only slightly across species in our lineage, between the australopithecines and the first *Homo*. A radical, and sudden, increase in brain size as seen in the evolution of *Homo* from then on must have required an equally radical and sudden change in the caloric intake. One such way to achieve such a change—to obtain many more kilocalories in the same time—is well known to us, and there is good and still-growing evidence in the fossil record that it was indeed used by our ancestors as early as 1 or even 1.5 million years ago,[5] just at the time the brain size of humans started to increase rapidly, as illustrated in figure 11.2. It is the transformation of foodstuffs—predigestion outside of the body, really, before food reaches the mouth—that goes by the name of "cooking."

Cutting, bashing, crushing, and otherwise tenderizing foods prior to chewing are also "cooking" in the looser sense of preparing foodstuffs as opposed to eating them as is.[6] Cooking in this looser sense is something early *Homo* and even their hunter-gather ancestors already did as far back as 4 million years ago, with their flint stone tools—and something the ancestors of great apes never did. Modern-like hands capable of precision grips, which facilitate handling tools and using them to process food, are unambiguously evident by 2 million years ago[7] (which got that particular *Homo* the name *habilis*). Hunter-gatherers had probably, therefore, already cut themselves some slack in what regarded the time required to ingest enough calories, thus freeing up time to do more interesting things with the extra brain neurons they could now afford, such as

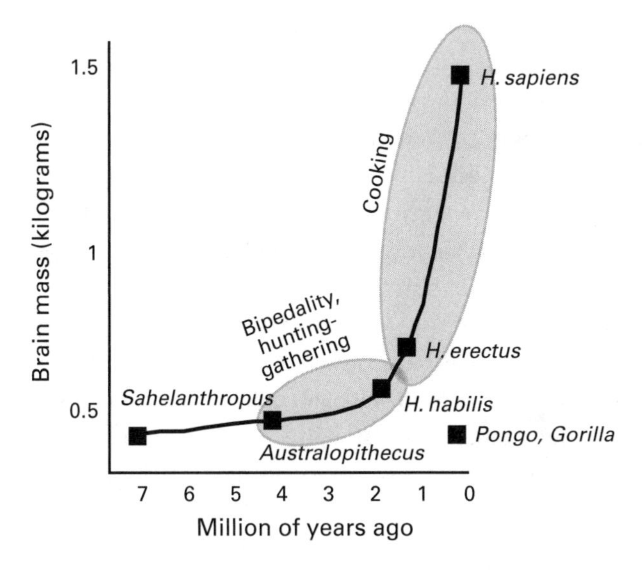

Figure 11.2
Rapid increase of brain mass in the *Homo* lineage in the last 1.5 million years coincides with the invention of cooking, probably by *Homo erectus*.

socializing and organizing hunts. But such primitive forms of cooking pale in comparison to the number of calories offered by the real deal: cooking with fire.

Homo culinarius

I had read Richard Wrangham's delicious book on the cooking hypothesis, *Catching Fire: How Cooking Made Us Human,*[8] in a single sitting during a long flight in late 2009, when a book publisher in Brazil asked me to evaluate it for the local readership. In a nutshell, the cooking hypothesis proposes that it was the invention of cooking by our direct ancestors and the resulting availability of cooked food that offered the larger caloric intake that allowed the brain of *Homo* to increase in size so rapidly in evolution. The circumstantial evidence of the drastic reduction in tooth and cranial bone mass, expected for a species that no longer had to use much effort to chew, was all there,* along with the fossil record that put the use of fire together with transforming foods between 1.0 and 1.5 million years ago. What Wrangham did not have then was an indication that cooking, or some other way to increase caloric yield from food, was not simply a bonus, an advantage for prehistoric *Homo,* but rather an essential requirement for their brains to become any larger.

At the time, my lab had already worked out that the human brain was just a scaled-up primate brain in its number of neurons, not an outlier, and I had strong suspicions that great apes, not us, were the outliers in the brain-body relationship—and for energy-related reasons, as we were able to demonstrate a few years later. But I was no primatologist, much less an anthropologist, and wasn't familiar with the field.

I was therefore delighted to find that Wrangham's thesis explained perfectly what we later confirmed: that once *Homo* were freed from the energetic constraint imposed by the raw food diet of all other animals, brain size increased rapidly—without humans ever stopping to be primates in the neuronal composition of their brains. Our ancestors never diverged from

*I also see the shortening of the intestines in modern *Homo* as a *consequence* of changing to the cooked food diet that allowed brain size to increase, rather than as a means to afford said larger brains, the central tenet of the expensive-tissue hypothesis (Aiello and Wheeler, 2006).

the same near-linear relationship that applies to all primates; they just managed to keep moving toward ever more neurons, well past the number other primates were and still are limited to by their raw food diets. In the years after 2009, and inspired by Wrangham's book, we pursued the hypothesis that the caloric intake from a raw food diet was so limiting that it curtailed any further increases in brain size in great apes—and ended up providing the numbers that showed, in black on white, that the availability of calories from the typical raw diet of primates is so limiting that, absent a way to overcome that limitation, the evolution of modern humans would simply not have been possible.[9] There was no way for the human brain to have emerged if not for a radical change in caloric intake. And the invention of cooking provided just that.

Cooking or no cooking, though, it is clear that the addition of massive numbers of neurons that we can infer to have driven the expansion of the brain in human evolution involved not only an increased caloric intake but also the adoption of a diet that consisted of softer foods. Evidence for that, reviewed in Wrangham's book, is the drastic change in the shape of the ridges on the skull that serve as attachments for the facial muscles required for chewing: the sagittal crest on top of the cranium and the zygomatic eminences on the cheeks, still so evident in gorillas and other great apes, disappeared in our lineage, while molar and canine teeth were radically reduced in size, even as the size of the brain cavity more than doubled within a single species, *Homo erectus*. Given that the shape of bones in the body is directly related to the force imposed on them by the attached muscles, the reduced size or even disappearance of bone features associated with effortful chewing indicates that those later *Homo erectus* populations no longer needed to use the brute force of their jaws and teeth to eat. And that is just what would be expected to happen over many generations for a species that moved to a habitually cooked diet.

Why Does Cooking Provide So Much More Energy?

Cooking, in the looser sense, is the transformation of food by any means before it enters the digestive tract—that is, the mouth. Cooking includes slicing, dicing, mincing, puréeing, mashing, seasoning, and marinating, all of which make food easier to chew and swallow. In the stricter sense, of course, cooking presupposes the use of heat to denature proteins, break carbohydrate chains, and otherwise modify the macromolecules of food, turning foodstuffs into smaller, softer, more easily chewable and enzymatically digestible versions of their former selves. Cooking with heat breaks down the collagen fibers that make meat tough and softens the hard walls of plant cells, exposing their stores of starch and fat. Cooked foods yield 100 percent of their caloric content to the digestive system because they are turned into mush inside the mouth, then digested completely by enzymes in the stomach and small intestine, where, once converted into amino acids, simple sugars, fatty acids, and glycerol, they are quickly absorbed into the bloodstream. In contrast, the same foods may yield as little as 33 percent of the energy in their chemical bonds when eaten raw because these harder foods are swallowed while still in pieces, and thus are broken down and digested only partially. Only the surface of the raw food crumbs is exposed to digestive enzymes in the stomach and small intestine; most of the unbroken starch finally gets digested in the large intestine by bacteria that keep the energy for themselves. From an energetic point of view, the main advantage of cooking foods is that, by making their digestion complete, it greatly increases their caloric yield.

The wonderful taste of cooked foods, of course, is another very important aspect of a cooked diet. It could be argued that it is an acquired taste, given that, in the wild, food has always been eaten raw in the nonhuman world. It was to examine this hypothesis that Richard Wrangham and his team offered fourteen chimpanzees a choice between raw and cooked carrots and sweet and white potatoes and found that six of them—nearly half—had an almost absolute preference for the cooked over the raw carrots and potatoes.[10] Later, Felix Warneken and Alexandra Rosati, also at Harvard University, found that chimpanzees were willing to return the raw food they were just given if they could exchange it for a cooked version—that is, they were willing to wait to have their food cooked.[11] Given the chance, even naive great apes prefer cooked foods—and so presumably did our ancestors. Dog owners who make the mistake of giving their pets cooked meals know that once the animals realize that there is cooked food in the world, it is hard to go back to kibble. An animal who enjoys the first taste of cooked food wants to have more, and it is easy to imagine that our ancestors must have felt the same. A few years ago, I witnessed the preference of chimpanzees for cooked foods when I visited the Great

Apes Project facility at Sorocaba, in the state of São Paulo, which housed at the time forty-eight chimpanzees rescued from circuses, where they had been mistreated, or retrieved from families who had finally come to their senses and realized that chimpanzees are not pets to be kept at home. Hugely muscular, aggressive animals, the chimpanzees at the facility charged at me for no other reason than that I was a stranger. Even behind thick Plexiglas walls reinforced with steel bars, the sight of a charging chimpanzee was scary enough to make me move back, just in case. But at noon, when the owners of the facility approached the window at the feeding area carrying bowls of cooked food, the charging muscular creatures turned into docile animals, waiting as patiently as they could manage for the next forkful of spaghetti and meatballs, mouths open in expectancy like human babies at lunchtime.

Even as cooking increases the caloric yield of foodstuffs, it decreases the amount of *time* required to obtain all those calories—simply because far less chewing is required to turn cooked foods completely into mush soft enough to be readily swallowed. Eating meat, which once was credited for the evolutionary expansion of the human brain, is actually extremely hard when the meat is raw: a medium-rare 200-gram (half-pound) steak that disappears from a human plate in 15 minutes would take over an hour to be swallowed raw without mincing or cutting. Once foods are cooked, less time is required to eat them, and more time becomes available to do other things with all those neurons that are now more easily affordable. And once the energy afforded by cooking turns a larger number of neurons from being a liability into being an asset, it becomes easy to envision a rapidly ascending spiral where larger numbers of neurons are selected for, given the cognitive advantage conferred to those individuals who have them and who now also have the time available to use them to hunt in groups, navigate the environment, look for better homing, hunting, and gathering grounds, and care for the well-being of their group, protecting it and passing on knowledge about where to find food and shelter.

To make a 2-million-year story short, I believe that the invention of cooking provides the simplest explanation for how humans went, in very little time, from being limited to eating raw foods, like any other animal, to, thanks to their now affordable larger number of neurons, planting their own food, collecting and distributing it in markets, growing entire civilizations around them, creating food distribution chains, supermarkets, electricity, refrigerators, and "industrialized foods," whether canned, frozen, or freeze-dried, that can be stored indefinitely and are readily available for consumption. There were lows in the process, of course; famines were paradoxically common in the recently industrialized world, in part because the very move toward agriculture, which anthropologist Jared Diamond calls "the worst mistake in the history of the human race,"[12] reduced the diversity of foodstuffs and thus placed humans at the mercy of blights, droughts, and wars. At the other extreme, "industrialized cooking," yet another invention of our neuron-rich brain, has taken so much water out of what we eat that it turned packaged ready-made foods into veritable caloric bombs that are hardly to be found in water-rich, home-cooked foods—which is why authors like Michael Pollan have defended a return to home cooking.

Because now, proud owners of energy-avid brains that don't seem to have caught onto the fact that energy is no longer a rare commodity, we overdo it. We humans are no longer at risk of not obtaining enough calories for our neuron-rich primate brains; we now suffer from being able to consume far too many calories. Ironically, given that it was our breaking away from the raw food diet of other primates that allowed us to become humans, we now resort to salads and other raw vegetables to remain so.

What about Modern Crudivores?

The very day that our study was published in 2012 showing that energy availability in a raw diet is so limiting that human evolution would not have been possible without a radical change in how humans got their calories such as the invention of cooking,[13] I received an e-mail explaining why our pretty graphs were wrong. There was no reason to think that cooking was necessary for a human diet, according to the writer, because "the caloric requirements of the human brain are trivial" and can be covered with "five bananas, thirteen oysters, seven turtle eggs or 69 cashew nuts."

The flaw in this argument is that *today* it may appear a trivial task to obtain the 500-plus kilocalories from raw foods to feed the human brain for a day, yes. But that is because the well-fed and hence enlarged brains of our ancestors came up with agriculture, food distribution schemes, and conservation technologies that make it possible to have refrigerators with fresh food stocked and readily available, without requiring extensive hours of foraging or collecting—not to mention the blenders and food processors that puree carrots and other hard-to-chew raw vegetables, staples of a modern human crudivore diet. But not having agriculture, food distribution, refrigerators, and grocery stores, how long would it have taken our ancestors to procure thirteen oysters or find a nest of turtle eggs?

Besides, despite the amenities of the modern, technological world, obtaining enough kilocalories from raw foods remains so difficult that the crudivore diet is *the* "tried-and-true" way to lose weight—although not without its drawbacks: the drastic weight loss that ensues, with a constant feeling of hunger, is often accompanied by malnourishment to the point that women on the diet stop menstruating. Karina Fonseca-Azevedo, the student who helped me work out the trade-off between number of neurons and body mass, tried a crudivore diet herself, in the early days of her Master's studies on the energetic constraints imposed by raw foods on other groups of mammals, and gave it up, after rapidly shedding 5 pounds. Meals now took forever, she told me—but really, she just couldn't bear the sight of yet another raw carrot.

What Is the Human Advantage, Then?

So what do we have that no other animal has and that explains our cognitive advantage? An outstandingly large number of cortical neurons, I say—even if we came to have them the primate way, without breaking any biological or evolutionary rules.* And what do we do that no other animal does and that allowed us to become humans? Forget about deceiving, reasoning, planning, counting, using language—other animals can also do those things, at least to some extent. We cook what we eat: this is the exclusively human activity, and one that allowed us to jump over

*No, I'm not saying that gaining more neurons thanks to cooking was the *only* change that occurred in human evolution; there is abundant evidence of small and large genetic changes of consequence to human anatomy and physiology. What I am arguing is that tripling the number of neurons in our brain and cramming in it the largest number of neurons found in the cerebral cortex of any species is the simplest, most basic and yet profound change that underlies our human advantage.

the energetic wall that still curbs the evolution of all other species and put us on a different evolutionary path from all other animals. As Michael Pollan told *Smithsonian* magazine, "Claude Lévi-Strauss, Brillat-Savarin treated cooking as a metaphor for culture, but if Wrangham is right, it's not a metaphor, it's a precondition."[14] So let's thank our *Homo culinarius* ancestors for our neurons, and show due respect for cooking. I know I now do.

12 ... But Plenty of Neurons Aren't Enough

Whenever I give talks about the fundamental role of cooking in human evolution, I know to expect the same two-part question: Why didn't the ancestors of modern great apes invent cooking as well? Couldn't a larger brain with more neurons have come first, and *that* allowed our ancestors to invent cooking?

As we saw in chapter 11, much had already happened to our cooking ancestors that differentiated them from the ancestors of chimpanzees and other great apes, and that probably made the invention of cooking more likely. Before cooking with fire was invented, our ancestors, and nobody else's, already benefited from the greater availability of calories from the increased range of hunting and gathering made possible by bipedality and the use of stone tools to chop and mince food. By the same token, our ancestors, and they alone, must have had a heightened demand for cognitive abilities and the opportunity to use those abilities both in hunting and in maintaining a tight social organization of now much more mobile bipedal individuals.

With some 30 billion neurons, modern great apes probably have about as many neurons in their brains as our exclusive hominin ancestor did 4 million years ago, and we know that 30 billion neurons are enough to use tools, such as sticks and stones—and fire. But that australopithecines, the first hominin species to become bipedal and use tools 4 million years ago, had a similar brain size and thus probably a similar number of neurons to noncooking modern chimpanzees and their ancestors illustrates a very important point. Having enough neurons is a *necessary* condition for complex and flexible behaviors, such as learning how to use fire and other tools to modify foods. But it is not a *sufficient* condition for behaviors to become more complex and flexible to the point of making even larger numbers of neurons an advantage that leads to ever more complex and flexible behaviors, in an ascending, self-reinforcing spiral that puts brains on the evolutionary road to expansion.

The difference is that having enough neurons endows a brain with the *capacities* for complex cognition, but turning those capacities into actual *abilities* takes a lifetime, if not generations, of learning, during which time the abilities developed are passed on and accumulated. This is the case to this very day in our modern, complex, busy lives. Those pieces of science writing which hail the human brain as a wonder often forget to mention that it never, *ever* starts life as a wonder, but rather as a 300-gram mass that doesn't do much yet—though it certainly holds great promise. The difference between capacities and abilities is made painfully clear in the extreme cases of children who are raised in near solitude, or with hardly any opportunities for learning: their brains remain arrested at the stage of promising raw material that could achieve much—but doesn't. In far less extreme cases, although we all supposedly have similar numbers of neurons in our brains, some of us in urban centers, exposed to schooling and challenges, use our brains to design and build skyscrapers and computers and to write books about the world and even our own brains, whereas others of us, raised in small groups in remote locations, use our brains mostly to build simple housing, to hunt, and to gather knowledge about the plants we collect. It's not enough to have plenty of neurons: they endow our brains with capacities, but not abilities.

Similarly, judging from brain size, we modern humans probably have just as many neurons nowadays as we had 200,000 years ago, and thus presumably similar cognitive capacities. Our cognitive *abilities*, on the other hand, are much more recent and are still growing very rapidly. They depend on our cognitive capacities just as technologies depend on available materials, building on them. And just as technologies also *generate* new materials, so our newly developed cognitive abilities change the ways our brains work, even if they don't add more neurons to them: just think of how learning to read modifies pattern recognition in the brain and opens up a whole new world of both possibilities and problems all at once. Our growing cognitive abilities develop technologies that expand the materials available, which in turn allow for new technologies to develop, placing an ever growing demand on our mental abilities. It's an ascending, self-reinforcing spiral--and it all begins with having enough neurons in the cerebral cortex, as we'll explore next.

More Neurons as Prime Material for Cognition

Every single item that we use in our lives, from simple pencils to complex fancy computer hardware, is made out of something that came from the ground. All materials originate in nature; what we do is transform them from their original incarnation to another, as a new material. This is a powerful realization that makes the human place in nature all the more remarkable: we are the species that transforms natural materials into other, more elaborate and flexible materials, with ever-growing possibilities.

It can be argued that having a large enough brain, with enough neurons in the cerebral cortex, must be a necessary condition for an animal to go from simply using materials made by nature to developing its own out of natural raw materials, for only humans have been able to do so. Neurons, as the information-processing units of the brain, can be thought of as raw material for cognition, much like Lego building blocks: a larger pile of Legos expands the number of possible structures that can be built with them. But just as there's always the chance they will remain just a pile of Legos if there is no use for them, so having enough neurons to support complex behaviors is no guarantee that they will be put to use in transformative ways.

Think of the history of materials available to and made available by us humans, modern and otherwise. We start with wood sticks and stone; expand to fire and what fire can help make available—cooked food, glass, molten metals and alloys; then, with the advent of agriculture around 10,000 years ago, we begin to multiply and diversify the materials we make available in a rapidly growing curve, shown in figure 12.1, that bears a striking resemblance to that of that of sudden and rapid growth of the human brain over evolutionary time.

Just as a larger brain paved the way to even larger brains, in a way that became quite rapid once larger brains were actually useful and no longer

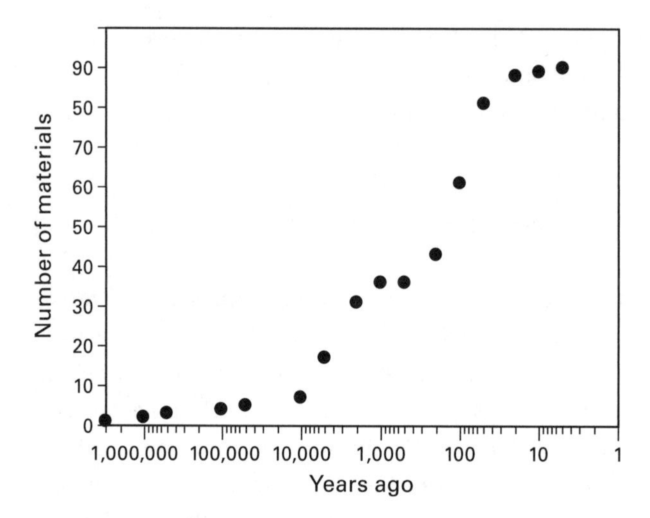

Figure 12.1
Total number of materials (of the first ninety that come to mind) available to humans or developed by humans over time.* Notice that with the advent of agriculture, 10,000 years ago, the number of materials available started to increase rapidly—just as brain mass did around 1.5 million years ago.

dangerously let alone prohibitively expensive, so the addition of new materials to the human toolkit seems to have fostered the addition of still other materials in greater and greater numbers—but only after agriculture created a demand for new solutions. Modern humans have been around for 200,000 years, but it was only 10,000 years ago, after the Ice Age (from 110,000 to 11,700 years ago) had finally ended, that agriculture was invented—and with it emerged a whole new world of possibilities to

*From oldest to most recent: stone, fire, obsidian, wood, bone, ceramics, copper, plaster, bronze, asphalt, cotton fabric, silk, gold, silver, graphite, glass, terracotta, parchment, papyrus, iron, latex rubber, wool, ivory, steel, pewter, marble, concrete, alum salts, paper, blown glass, naphtha, coke (fuel), petroleum oil, porcelain, gunpowder, dental amalgam, glass lenses, mirrors, molten glass, Portland cement, tungsten, ,cardboard, aluminum, sheet glass, creosote, rubber, kerosene, vinyl, tanned leather, plastic, TNT, dynamite, gasoline, celluloid, asphalt concrete, PVC, photovoltaic cells, dental ceramics, silicon carbide, silicone, bulletproof glass, cellophane, Bakelite, stainless steel, Pyrex, metal foam, hiduminium alloy, neoprene, polystyrene, Plexiglas, nylon, aerogel, guncotton, titanium, Teflon, PET, synthetic diamonds, Styrofoam, ANFO, carbon fibers, Lycra, graphene, Kevlar, liquid crystals, optical fibers, Twaron, fullerene, SEAgel.

be pursued and problems to be solved. It makes sense: with new materials, new possibilities emerge, leading to steplike changes in the realm of the possible, but only if the changes are fostered by new needs, such as those brought by the invention of agriculture and the novel problems of harvesting, food storage, and crop management. These means and opportunities to use new materials, creating new possibilities, are technological innovations: new objects, processes, and systems that provide new and better solutions to preexisting problems, and may even create new problems to be solved. Sticks and mud may have been enough to solve the early problem of building shelter where nature provided none, but that solution must soon have led to the realization that stick housing was not strong or resistant enough to last very long. The invention of clay bricks solved that problem—but brick buildings become unstable and collapse past a certain height, a problem that iron and steel solved. Later, sheet glass in combination with structural steel took building to a whole new level, allowing skyscrapers to completely change the skyline of modern cities—and creating a new problem: competing for the status of raising the tallest gravity-defying structure. A previous incarnation of sheet glass, molten glass, had led to the invention of lenses, mirrors, and photography, but for the latter, it posed the problem of making the first cameras heavy and cumbersome—a problem solved by the invention of celluloid film, ushering in the technology of motion pictures and a whole new industry, cinema, with its own characteristic set of new problems to be solved. Celluloid itself was soon replaced by other plastics, and one of them, vinyl, in the form of phonograph records, went on to make music a household item, rather than an elite and expensive cultural happening, which, in turn, spawned a new class of elite individuals: rock stars. Music, however, only became the complex, multiphonic amalgam of sounds we are now used to thanks the invention of a new material, paper, and of two new processes: musical notation, developed around 1000 AD, without which the playing and transmitting of music was limited by people's memories and voices; and Gutenberg's moveable-type printing press, in 1450. The resulting new problem of carrying and handling bulky, heavy books of printed music and printed anything else was then addressed by further technologies developed in the twentieth century, all built on new silicon-based materials, which took printing to another level and medium. As a result, now I can consult treatises on history, such as E. H.

Gombrich's *A Little History of the World* (2005), one of my favorite books ever; on the history of materials, such as Mark Miodownik's *Stuff Matters* (2014); and on the history of music, such as Howard Goodall's *The Story of Music* (2014), to write this chapter, and a new, electronic copy of Wrangham's *Catching Fire* (bought upon finding that my dad, in good family tradition, had "borrowed" my printed copy the last time he visited) to double-check information, and access a whole body of knowledge recorded in what we still call "books"—all in the easy-to-carry, convenient form of a light black plastic tablet. And, true to form, this new technological solution lets me worry about the next problem: what to do when the tablet breaks (as it has all too often).*

Although technologies rely on new materials and often lead to their development to meet new technological requirements, some materials are developed without a use in sight. Thus the superstrong, superconductive, nearly transparent form of carbon known as "graphene" is a brand-new material for which not a single practical use is known today—although, just as with basic science itself, a number of possibilities can be expected in the near future. New materials make new technological leaps possible, just as more neurons supposedly make new cognitive leaps possible. New technologies, in turn, lead to the development of new materials, which allow more new technologies, in an upward, self-feeding spiral. And so it went in our early modern history: from wood and stone tools to ropes and metals, ceramic and glass, to textiles and paper, to concrete, rubber, and blown glass, all made available around 2,000 years ago.

But then, about 2,000 years ago, that spiral stopped self-feeding anymore, as indicated by the saddle in the slope in figure 12.1, and shown more clearly in a plot of when new materials were developed, in figure 12.2. With a brief respite a bit before 1,000 years ago, when the Chinese

*"Clouds" theoretically solve that problem; one can "always" reconnect and recover the tablet's content—provided electricity and Internet connectivity are restored. Maybe because I live in a third world country, I don't take those as givens—which is why I keep printed copies of my favorite books. Technology is great, but having direct access to knowledge is more important, and there still is nothing like having to rely on my neurons alone to obtain it—well, that and on stored paper, even though that does tend to disintegrate with time. Yes, problems in need of solutions are never ending, and therefore so is the drive for new technologies. It all starts with having enough neurons to appreciate the problem.

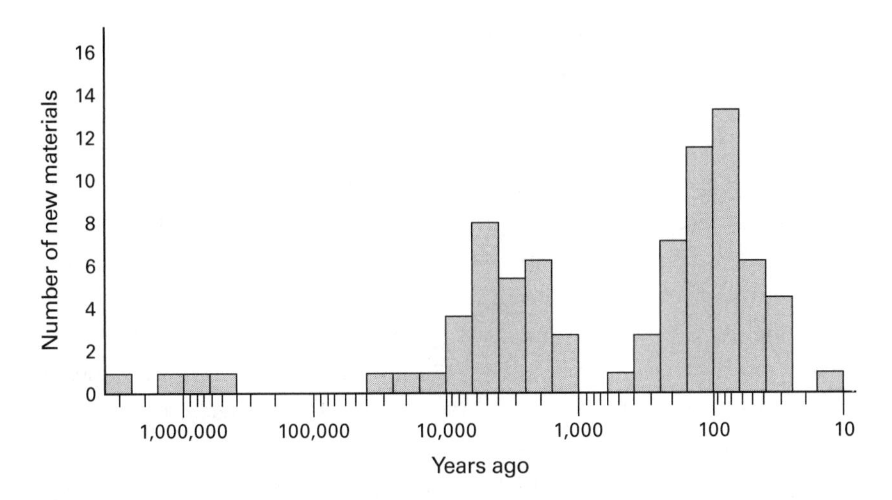

Figure 12.2
There is a gap in the number of materials (of the first ninety that come to mind) newly available to humans or developed by humans around 1,000 years ago—but starting some 400 years ago, new materials once again became available rapidly. Although this involves only the first ninety materials I could think of, it shows an important difference in their dates of creation: The history of development of new materials has not been a linear, progressive one.*

developed porcelain and gunpowder, the development of new materials by humanity came to a halt around 1,000 years ago, only to start up again, slowly, some 600 years later, with the invention of glass lenses, molten glass, and mirrors, and then once more around 150 years ago, with invention of cardboard, the first polymers (such as celluloid), Portland cement and asphalt concrete, and then again some 80 years ago, with the second generation of polymers (nylon, polystyrene), Pyrex and Plexiglas, and new metallic alloys.

Why, around 1,000 years ago, was having agriculture and new materials to work with suddenly no longer enough to foster the appearance of ever more new materials? What happened between 2,000 years and 300 years ago that could have had put such a powerful brake on progress in the development of new materials? It was a period of our history aptly described by two words: "Dark Ages."

**Stuff Matters*, a delightful book by Mark Miodownik (2014), explores in much greater detail the influence that materials have had in human history.

Technology, Culture, and the Downfall of the Neanderthals

The ability to develop new technologies to use available materials, to develop new ones and to pass both new technologies and new materials on was all but lost during the Dark Ages, between the collapse of the Roman Empire in 476 AD and the Renaissance in the early fifteenth century. During the Dark Ages, very few people could read or write, and knowledge gave way to superstition. The human brain still had presumably the same number of neurons in the cerebral cortex, and the same basic materials were still available. Nevertheless, without the means to explore those materials and pass on the technologies, the upward spiral of ever more new materials and ever more new technologies was interrupted. Without cultural transmission, technology dies in one generation.

Cultural transmission must have been equally fundamental for the evolution of human abilities, once the increased capacities were afforded by our newly achieved and outstanding number of cortical neurons. We can think of our brains, and our cortical neurons in particular, as biological materials to work with. How we use these biological materials in our heads, and therefore what they can produce at any moment in time, depends on the available mental technologies—which include mental processes and systems to solve problems and come up with new ones, starting with reading and arithmetic. In turn, whether and how those mental technologies thrive and progress and even allow the development of new technologies themselves, eventually expanding to the development of new materials and new technologies with which to transform them, depends on cultural transmission to the new generations—which, of course, depends on having the required neurons to teach and learn.

Having enough cortical neurons to support greater cognitive capacities, once these were proven to be affordable, can thus be seen as a necessary but not sufficient step toward acquiring the cognitive abilities that, once proven beneficial, were selected for, leading to still more cortical neurons and generating the upward spiral of increasing brain size in human evolution. And this happened only in our human lineage because only our human ancestors possessed the required technology and culture to move the process along.

This is not to say that cultural transmission is something other primates and other animals lack. They don't. Among bottlenose dolphins, for example, there are "spongers," who wear sponges on their noses that protect them from sharp objects and stingray barbs and allow those dolphins to explore new foraging niches; this "sponging" behavior is passed on from mothers to offspring.[1] Similarly, chimpanzees of different clans and in different geographical areas make their own particular use of sticks and stones as tools and pass that use along to their offspring,[2] although these tools are not modified as ours were 3.4 million years ago. They have the culture—but they don't have the technology.

Thus, in our technological history, square one, not only for all primates but also for many other animals, is the use of natural materials as tools. Chimps do it, monkeys do it, even birds do it.* The ability to use natural materials as tools is widespread, if not universal, among vertebrates and at least among some invertebrates, such as ants and octopodes. Then came what I will call the "First Technological Revolution": *making* tools—but, within primates, it came only to those who had the requisite number of cortical neurons, the helpful anatomical transformations (shorter, stronger fingers and longer opposable thumbs, most notably), and the hunter-gathering culture that made tools all the more valuable. Some 4 to 3.4 million years ago, our australopithecine ancestors had already diverged from our common lineage with chimpanzees and—well after the divergence of the gorilla and orangutan lineages—learned to develop stone tools through the modification of naturally occurring stones. That was the beginning of the Stone Age, almost 2 million years before the beginning of the rapid upward slope in the brain size of *Homo erectus* that can be traced as part of our exclusive evolutionary history. Not only could our ancestors of some 4 million years ago use tools, but, being bipedal, they could also carry them as they roamed farther and farther afield.

The "Second Technological Revolution" in human evolutionary history, learning the controlled use of fire around 1.5 million years ago to cook

*It seems fitting that, as we now know, birds such as corvids and parrots have even more neurons in their telencephalon than primates with cortices of similar mass. Indeed, a raven has as many neurons in its telencephalon as a macaque has in its much larger cortex. But that, too, is a whole other story.

foods, thus happened not simply to a primate species that had enough neurons in its cerebral cortex to plan and execute the transformation of natural materials into tools, but to a species that already had that capacity *plus* a budding technological and cultural history and a whole new level of problems to solve—navigating an expanded environmental range, coordinating groups to hunt, relying increasingly on communication—that set that species apart from other equally large-brained primates of the time. With stone technology literally in hand, and with the newly acquired technology of cooking, our ancestors of some 1.5 million years ago passed valuable technological knowledge on to the new generations through cultural transmission. It is in this context that I answer the second most common question I get at talks to the general public: What would happen if we started feeding cooked food to great apes (much as they've been doing at the Great Apes Project sanctuary in Sorocaba for several years now)? The first part of my answer, only partly tongue in cheek, is "Ask me again in one million years"—for evolutionary change takes generations to become obvious. It does create fat animals, though. Ingesting more calories rapidly increases the waistline, but doesn't make the adult brain immediately grow any larger. Much to the contrary, there is evidence that being overweight can accelerate age-related brain shrinkage in humans.[3] The effect of having more calories on evolving larger brains must occur through natural selection acting during gestation and in early life, when neurons are being added to and removed from the brain, and therefore can only appear in a population over several generations, as more and more individuals with more neurons and larger brains survive. The second part of my answer, however, is "There's a clear difference between a species that already has the technological and cultural means to cook by itself, and another that certainly appreciates cooked foods when they're offered, but that is still far from knowing how to cook for itself." The great apes who receive cooked food today are *not* at the same starting point where our bipedal, hunter-gatherer ancestors were 1.5 million years ago, when they presumably learned to cook with fire and their brains began to grow rapidly over the generations. The problems that human ancestors and modern great apes have to solve are very different in complexity, and it is the species with the hardest problems to solve that should benefit the most from affording more neurons. We humans have been feeding cooked food to our dogs, and to other domesticated animals, for centuries—with no sign that any are the smarter for it. At our homes, a dog's life is not really that cognitively demanding.

Homo sapiens is not the only recent variety of humans ever to cook, though. *Homo neanderthalensis*, our cousins who made it much earlier to Europe than modern humans, shared with us a common ancestor who lived more than 400,000 years ago. Neanderthals had therefore already inherited the same traditions of tool making, hunting large animals using spears, making fires, and cooking their food by the time they wandered off to Europe, while *sapiens* stayed behind in southern Africa. Why, then, did only the *sapiens* variety of cooks survive?

No one knows for certain, but hypotheses abound, including that modern humans simply outbred the Neanderthals or that they directly confronted them and killed them off in one fell swoop (although evidence of interbreeding argues against this).* I favor an explanation akin to what happened much later, when newcomers and indigenous peoples met again, now in the Americas, at the turn of the sixteenth century: the subjugation over centuries of one variety of humans by another, biologically equivalent but much more powerful in technology and culture. The confrontation of Cavalry or Conquistadores versus Indians in the Americas is, to my mind, a repetition of what must have been the confrontation of incoming *sapiens* versus resident *neanderthalensis* much earlier in Europe. In both cases, the confronting populations were variants of the same species, for the confrontations resulted in mixed-race offspring; in both cases, populations were meeting again after their ancestors had gone their separate ways (Neanderthals and modern humans from southern Africa meeting again in Europe; then *sapiens* from Europe and *sapiens* from Asia meeting again in the Americas once sailboats were invented). And, in both cases, the incoming population devastated the natives—because, over time, the two diverging populations had grown apart so much in technological and cultural achievements, that it was nearly inevitable

*This very interbreeding, by the way, argues that *Homo sapiens* and *Homo neanderthalensis* were one and the same species, at least according to the biological concept of species—which is why I refer to them as "varieties" of humans, not different "species," despite their Latin (Linnaean) names.

that the incoming would overpower the residents whenever and wherever they met.

Indeed, that confrontation of Europeans versus American Indians in the early sixteenth century almost certainly recaptured what had already happened between their ancestors some 60,000 years before. The two diverging populations of *neanderthalensis* and *sapiens* humans that met in Europe between 60,000 and 50,000 years ago had grown apart both physically and culturally, coming to differ greatly in their achievements. By 70,000 years ago, early *sapiens* humans in Africa were trading over long distances, making new tools, including stone arrowheads and bone tools such as harpoons for fishing, and already had symbolic art.[4] By 50,000 years ago, *sapiens* humans had developed new technologies that allowed them to create more versatile stone and bone tools for making clothing, nets, lamps, fishhooks, and even flutes, and to build semipermanent housing.

In contrast, *neanderthalensis* humans, who evolved in colder climates, were more robust than their tropical *sapiens* cousins—although, despite popular claims to the contrary, brain mass varied over a similar range in both.[5] But *neanderthalensis* humans in Europe had very little in terms of technological innovations or symbolic art. And, in addition to their much richer technological culture, the incoming *sapiens* humans probably had clearer, easier-to-interpret speech than *neanderthalensis* humans, due to anatomical differences in the speech apparatus,[6] even though both had the same FoxP2 gene variant whose expression in the brain is associated with the evolution of human speech.[7]

So, around 40,000 years ago, the *neanderthalensis* variety of humans disappeared, overpowered, one way or another, by the *sapiens* variety[8]—although some *neanderthalensis* genes remain, having been incorporated into the *sapiens* genome some 50,000–60,000 years ago,[9] the undisputable evidence of our intermixing. And then there was one: *Homo sapiens*.

The Modern Technological Revolutions

After the end of the Ice Age, around 12,000 years ago, human populations in different parts of the world began what would later prove to be an irreversible process: settling down, exploring the land in the vicinity, and eventually farming the land. Whereas one hunter-gatherer couple can manage to collect between 5,000 and 8,000 kilocalories per day,* one farming couple can harvest around 13,000 kilocalories per day, and more easily afford a family. Farming thus provided more energy from the same land, and allowed an unprecedented population boom that could not have been sustained by hunting and gathering. But population growth came at a price: the nutritional stress associated with a poorer and less varied diet and their settled lifestyle led to the appearance among farmers of malnutrition and infectious diseases that were rare or absent in hunter-gatherers. While the downside of agriculture may be evident in hindsight, once farming afforded larger populations that could no longer be sustained by hunting and gathering, there was no going back.[10]

Farming also brought other advantages. With all its new problems related to plowing the land, irrigating it, harvesting the crops, and then distributing the harvest, farming created a whole new set of problems that the human brain, with its large number of cortical neurons, was apt to solve. In what I will call the "Third Technological Revolution," agricultural humans developed the tools not just to control their environment and their crops, but also to modify them (for most of the modern edible plants are distant versions of their original selves, starting with corn but including carrots, tomatoes, and wheat; modern genetic engineering is but an accelerated version of what we humans have been doing over the last 10,000 years). Agriculture gave rise to modern civilizations and the hierarchical social structures that persist to this day. New problems led to new technologies, which led to new materials, which allowed solving old problems and creating new ones. Civilizations blossomed, in a part of our history well told in Tom Standage's aptly named book *An Edible History of Humanity* (2009).

Made possible by the scientific revolution of the renaissance, when systematic inquiry started replacing dogmas about the unknown, the "Fourth Technological Revolution" is far more recent: the Industrial Revolution of the nineteenth century that modernized farming itself through the invention of human-operated machines that mechanized agricultural work. As had happened when manual farming was invented, the Industrial

*These many calories would not have been available to the ancestral hunter-gatherers of 4 to 2 million years ago; without cooking, they would only have absorbed part of the nutrients from the raw foodstuffs. In the modern world, hunter-gatherers cook, so all of these calories are effectively transferred to the body.

Revolution led to a new round of population growth, afforded by the availability of even larger amounts of food.

Much more recently, in the "Fifth Technological Revolution," we went beyond machines operated by humans and created *automated* machines that replace human work with the turn of a button. In my favorite TED talk ever, the Swedish physician and statistician Hans Rosling explains why he considers the washing machine the most important piece of modern technology: because it changed the lives of women in the twentieth century, and still does, by making time available for women to become educated and participate more closely in the education of their children.* Again and again, we see the same motif: technology, be it through cooking, farming, cranking levers, or pushing buttons, makes time available by solving one type of problem—only to create new problems, which in turn foster new technologies. And so we find ourselves in the midst of the "Sixth Technological Revolution": outsourcing to machines not just our physical work, but our mental abilities, too. Memory and cognition, to many, have become optional, outsourced to their cell phones and Internet browsers.

This is a serious problem that we tend to ignore—and one that makes fictional accounts of catastrophic scenarios so appealing to me.† The problem is that we have reached a point where too few of us dominate the current technologies. Who among us would know how to melt and work metal, much less use it to build a car, telephone, or computer from scratch? Being able to call myself a scientist doesn't even guarantee that I would know how to craft a simple pencil. So much of modern technology is no longer within the grasp of any single individual. We boast of having come such a long way since the ancient Greeks—but, unlike them, we can no longer be experts in architecture, biology, and physics all at the same time. That is why science (the knowledge) and engineering (the crafts) must be carefully cultivated, documented, and passed on to the next generations. It

*See Hans Rosling, "The Magic Washing Machine" at http://www.ted.com/talks/hans_rosling_ and_the_magic_washing_machine?language=en.

†For instance, one of my recent postapocalyptic favorites is *Station Eleven*, a novel by Emily St. John Mandel (2014) that explores the workings of an early postpandemic world without electricity, in which airports become safe outposts due to the abundance of stored foods, natural daylight, and toilets.

is not enough to have a remarkable number of cortical neurons to achieve remarkable feats: we stand on the shoulders of all those who came before us—and, by now, the achievements of our species as a whole far surpass those of any one individual. Humankind has long transcended man. It is the self-reinforcing pairing of technological innovations and cultural transmission, made possible by the outstanding number of neurons in our cerebral cortex, that shaped our capabilities into abilities and got us where we are—for better or for worse.

Epilogue: Our Place in Nature

As it turns out, there *is* a simple explanation for how the human brain, and it alone, can be at the same time similar to others in its evolutionary constraints, and yet so different to the point of endowing us with the ability to ponder our own material and metaphysical origins. First, we are primates, and this bestows upon humans the advantage of a large number of neurons packed into a small cerebral cortex. And second, thanks to a technological innovation introduced by our ancestors, we escaped the energetic constraint that limits all other animals to the smaller number of cortical neurons that can be afforded by a raw diet in the wild.

So what do we have that no other animal has? A remarkable number of neurons in the cerebral cortex, the largest around, attainable by no other species, I say. And what do we do that absolutely no other animal does, and which I believe allowed us to amass that remarkable number of neurons in the first place? We cook our food. The rest—all the technological innovations made possible by that outstanding number of neurons in our cerebral cortex, and the ensuing cultural transmission of those innovations that has kept the spiral that turns capacities into abilities moving upward—is history.

And so we flourished: over the last 200,000 years, our large brain with its highly neuron rich cerebral cortex (but all the while a perfectly normal primate cortex) invented culture, agriculture, civilization, markets, supermarkets, electricity, supply chains, refrigerators—all those things which conspired to make countless calories now easily available. So much so that the 2,000 kilocalories we require for the day can be consumed in a single sitting at our favorite fast-food place around the corner. No hunting or gathering, planting or harvesting is required. Not even cooking is required anymore, at least not by ourselves: our technology-savvy civilization allows outsourcing—even of our own cognition, if need be.

Our work on the human brain[1] was published in the year of Darwin's 200th anniversary and the 150th anniversary of the publication of his seminal book *On the Origin of Species* (1859). In talks to the general public, I always show a picture of the blue whale–sized mammal that would hold an implausible rodent-sized, 36-kilogram (80-pound) brain containing our 86 billion neurons, according to the rules that describe how nonprimate brains are put together. I then contrast it with what a generic primate with 86 billion neurons would look like, according to the scaling rules that we uncovered: a 66-kilogram (150-pound) animal with a 1,240-gram (2.75-pound) brain, which I illustrate with a well-known portrait of Darwin with "his brain" exposed by transparency. "So Darwin was a primate," I conclude, "and so am I, and so are each of you in the audience." To my surprise—and, I confess, initial disappointment—I always see smiling faces nodding placidly in the audience.

As a biologist, I am flattered and honored to be the one to present Darwin with posthumous evidence that, as he himself claimed, we were created in the image of other primates (I like to think that he would have been pleased with that finding). Once my initial disappointment wears off (I still expect protests! Disbelief! Skepticism!), it's a relief to see that the general public in the twenty-first century takes being called "a primate" so well. We've come a long way since Darwin, thanks largely to his work, which paved the way toward a better understanding of our place on Earth. We may be the species that amasses the largest number of neurons in its cerebral cortex, which makes us unique in that way. But we got here thanks to contingencies that put enough technology in the hands of our ancestors, which secured them enough access to food and the ability to transform it that allowed them to overcome the energetic constraints that still apply to all other animals on Earth. We jumped over that energetic wall and grew to invent human-operated machines, self-operating machines, and even machines that can replace ourselves, or at least our cognitive selves. But we never stopped being primates.

Appendixes

Appendix A: Body Mass, Brain Mass, and Number of Neurons

Values are listed for each species in kilograms (kg), grams (g), millions (M), or billions (B). Species are listed in order of increasing brain mass. All values refer to the whole brain (without olfactory bulbs). Detailed information and other values can be found in Herculano-Houzel, Catania, Kaas, and Manger, 2015.

Species	Body mass	Brain mass	Brain neurons
Smoky shrew (eulipotyphlan)	7.8 g	0.176 g	36 M (10 M in cerebral cortex)
Short-tailed shrew (eulipotyphlan)	16.2 g	0.347 g	55 M (12 M in cerebral cortex)
Mouse (rodent)	40.4 g	0.402 g	68 M (14 M in cerebral cortex)
Hairy-tailed mole (eulipotyphlan)	42.7 g	0.759 g	124 M (16 M in cerebral cortex)
Star-nosed mole (eulipotyphlan)	41.1 g	0.802 g	131 M (17 M in cerebral cortex)
Golden mole (afrotherian)	79.0 g	0.812 g	65 M (22 M in cerebral cortex)
Golden hamster (rodent)	168.1 g	0.965 g	84 M (17 M in cerebral cortex)
Eastern mole (eulipotyphlan)	95.3 g	0.999 g	204 M (27 M in cerebral cortex)
Elephant shrew (afrotherian)	45.1 g	1.040 g	129 M (26 M in cerebral cortex)
Rat (rodent)	315.1 g	1.724 g	189 M (31 M in cerebral cortex)
Mouse lemur (primate)	60.0 g	1.799 g	155 M (22 M in cerebral cortex)

Appendix A: (continued)

Species	Body mass	Brain mass	Brain neurons
Spiny rat (rodent)	223.5 g	2.078 g	202 M (26 M in cerebral cortex)
Four-toed elephant shrew (afrotherian)	132.5 g	2.440 g	157 M (34 M in cerebral cortex)
Tree shrew (scandentian)	172.5 g	2.752 g	261 M (60 M in cerebral cortex)
Guinea pig (rodent)	311.0	3.656 g	234 M (44 M in cerebral cortex)
Prairie dog (rodent)	1.5 kg	5.321 g	438 M (54 M in cerebral cortex)
Grey squirrel (rodent)	500 g	5.548 g	454 M (77 M in cerebral cortex)
Marmoset (primate)	361.0 g	7.780 g	636 M (245 M in cerebral cortex)
Rabbit (lagomorph; grouped with rodents under Glires)	4.6 kg	9.132 g	494 M (71 M in cerebral cortex)
Galago (primate)	946.7 g	10.150 g	936 M (226 M in cerebral cortex)
Tree hyrax (afrotherian)	1.2 kg	12.800 g	504 M (99 M in cerebral cortex)
Owl monkey (primate)	925 g	15.730 g	1.5 B (442 M in cerebral cortex)
Rock hyrax (afrotherian)	2.5 kg	16.853 g	756 M (198 M in cerebral cortex)
Agouti (rodent)	2.8 kg	17.628 g	795 M (111 M in the cerebral cortex)
Squirrel monkey (primate)	859 g	30.216 g	3.2 B (1.3 B in cerebral cortex)
Long-tailed monkey (primate)	5.7 kg	46.162 g	3.4 B (801 M in cerebral cortex)
Capuchin monkey (primate)	3.3 kg	52.208 g	3.7 B (1.1 B in cerebral cortex)
Bonnet monkey (primate)	8.0 kg	61.470 g	3.8 B (1.6 B in cerebral cortex)
Pig (artiodactyl)	100 kg	64.180 g	2.2 B (307 M in cerebral cortex))
Capybara (rodent)	47.5 kg	74.734 g	1.6 B (306 M in cerebral cortex)
Rhesus monkey (primate)	3.9 kg	87.346 g	6.4 B (1.7 B in cerebral cortex)
Springbok (artiodactyl)	25 kg	106.074 g	2.7 B (397 M in cerebral cortex)
Baboon (primate)	8 kg	151.194 g	10.9 B (2.9 B in cerebral cortex)
Blesbok (artiodactyl)	60 kg	154.718 g	3.0 B (571 M in cerebral cortex)
Greater kudu (artiodactyl)	218 kg	306.860 g	4.9 B (762 M in cerebral cortex)
Giraffe (artiodactyl)	470 kg	537.218 g	10.8 B (1.7 B in cerebral cortex)
Human (primate)	70 kg	1,509 g	86.1 B (16.3 B in cerebral cortex)
African elephant (afrotherian)	5,000 kg	4,619 g	257 B (5.6 B in cerebral cortex)

Appendix B: Scaling Rules

The full equations for the scaling rule functions are listed here, including r^2 and p-values.

Figure	Dependent variable	Independent variable	Species	Function	r^2	p-value
4.3	Brain mass (grams)	Brain neurons	Rodents (n = 6)	$M_{BR} = e^{-28.68255} N_{BR}^{1.550\pm0.106}$	0.982	0.0001
4.3	Brain mass (grams)	Brain neurons	Primates (n = 6) and scandentia (n = 1)	$M_{BR} = e^{-19.25873} N_{BR}^{1.046\pm0.062}$	0.982	< 0.0001
4.4	Mass of cerebral cortex, including white matter (grams)	Number of neurons in cerebral cortex	Rodents (n = 10)	$M_{CX} = e^{-29.36857} N_{CX}^{1.699\pm0.096}$	0.975	< 0.0001
4.4	Mass of cerebral cortex, including white matter (grams)	Number of neurons in cerebral cortex	Primates (n = 11, excluding humans), and scandentia (n = 1)	$M_{CX} = e^{-17.58313} N_{CX}^{1.014\pm0.070}$	0.954	< 0.0001
4.5	Mass of cerebellum, including white matter (grams)	Number of neurons in cerebellum	Rodents (n = 10)	$M_{CB} = e^{-26.46238} N_{CB}^{1.349\pm0.073}$	0.977	< 0.0001
4.5	Mass of cerebellum, including white matter (grams)	Number of neurons in cerebellum	Primates (n = 12, excluding humans), and scandentia (n = 1)	$M_{CB} = e^{-19.17512} N_{CB}^{0.956\pm0.036}$	0.985	< 0.0001
4.6	Mass of rest of brain (grams)	Number of neurons in rest of brain	Rodents (n = 10)	$M_{ROB} = e^{-26.20184} N_{ROB}^{1.568\pm0.252}$	0.829	0.0003
4.6	Mass of rest of brain (grams)	Number of neurons in rest of brain	Primates (n = 11, excluding humans), and scandentia (n = 1)	$M_{ROB} = e^{-18.73558} N_{ROB}^{1.126\pm0.148}$	0.853	< 0.0001

Appendix B: (continued)

Figure	Dependent variable	Independent variable	Species	Function	r^2	*p*-value
4.8 and 5.2	Mass of cerebral cortex, including white matter (grams)	Number of neurons in cerebral cortex	Rodents (n = 10), eulipotyphlans (n = 5), afrotherians (n = 6), artiodactyls (n = 4)	$M_{CX} = e^{-27.90849} N_{CX}^{1.612\pm0.038}$	0.987	< 0.0001
4.8 and 5.2	Mass of cerebral cortex, including white matter (grams)	Number of neurons in cerebral cortex	Primates (n = 11, excluding humans), and scandentia (n = 1)	$M_{CX} = e^{-17.58313} N_{CX}^{1.014\pm0.070}$	0.954	< 0.0001
4.10 and 5.3	Mass of cerebellum, including white matter (grams)	Number of neurons in cerebellum	Rodents (n = 10), afrotherians (n = 5), artiodactyls (n = 5)	$M_{CB} = e^{-25.11687} N_{CB}^{1.283\pm0.035}$	0.987	< 0.0001
4.10 and 5.3	Mass of cerebellum, including white matter (grams)	Number of neurons in cerebellum	Primates (n = 12, excluding humans), and scandentia (n = 1)	$M_{CB} = e^{-19.17512} N_{CB}^{0.956\pm0.036}$	0.985	< 0.0001
4.10 and 5.3	Mass of cerebellum, including white matter (grams)	Number of neurons in cerebellum	Eulipotyphlans (n = 5)	$M_{CB} = e^{-21.16534} N_{CB}^{1.028\pm0.084}$	0.980	0.0012
4.12 and 5.4	Mass of rest of brain (grams)	Number of neurons in rest of brain	Primates (n = 12, including humans)	$M_{ROB} = e^{-19.96155} N_{ROB}^{1.198\pm0.116}$	0.915	< 0.0001
4.12 and 5.4	Mass of rest of brain (grams)	Number of neurons in rest of brain	Afrotherians (n = 5), eulipotyphlans (n = 5), rodents (n = 10), artiodactyls (n = 5), scandentia (n = 1)	$M_{ROB} = e^{-32.08683} N_{ROB}^{1.917\pm0.118}$	0.916	< 0.0001

Appendix B: (continued)

Figure	Dependent variable	Independent variable	Species	Function	r^2	*p*-value
4.13	Neurons per milligram in cerebral cortex, including white matter	Number of neurons in cerebral cortex	Rodents (n = 10), eulipotyphlans (n = 5), afrotherians (n = 6), artiodactyls (n = 5)	$DN_{CX} = e^{20.979921} N_{CX}^{-0.610\pm0.037}$	0.918	< 0.0001
4.14	Neurons per milligram in cerebellum, including white matter	Number of neurons in cerebellum	Rodents (n = 10), afrotherians (n = 5), artiodactyls (n = 5)	$DN_{CB} = e^{18.200391} N_{CB}^{-0.283\pm0.035}$	0.784	< 0.0001
5.1	Brain mass (grams)	Brain neurons	Afrotherians (n = 5), eulipotyphlans (n = 5), rodents (n = 10), artiodactyls (n = 5), scandentia (n = 1)	$M_{BR} = e^{-27.27905} N_{BR}^{1.469\pm0.046}$	0.977	<0.0001
5.1	Brain mass (grams)	Brain neurons	Primates (n = 10, excluding humans)	$M_{BR} = e^{-21.59394} N_{BR}^{1.156\pm0.052}$	0.984	< 0.0001
7.1	Mass of cerebral cortex (grams)	Mass of rest of brain (grams)	Primates (n = 12, including humans)	$M_{CX} = e^{1.1367593} M_{ROB}^{1.294\pm0.069}$	0.972	< 0.0001
7.1	Mass of cerebral cortex (grams)	Mass of rest of brain (grams)	Afrotherians (n = 6), rodents (n = 10), eulipotyphlans (n = 5), artiodactyls (n = 5)	$M_{CX} = e^{0.4743052} M_{BD}^{1.179\pm0.023}$	0.991	< 0.0001

Appendix B: (continued)

Figure	Dependent variable	Independent variable	Species	Function	r^2	*p*-value
7.5	Number of neurons in cerebellum	Number of neurons in cerebral cortex	Afrotherians (n = 5), rodents (n = 10), eulipotyphlans (n = 5), artiodactyls (n = 5), primates (n = 12)	$N_{CB} = -4.133 \times 10^8 + 4.2\ N_{CX}$	0.985	< 0.0001
7.6	Number of neurons in cerebral cortex	Number of neurons in rest of brain	Primates (n = 12, including humans)	$N_{CX} = e^{-4.617208} N_{ROB}^{1.391 \pm 0.158}$	0.885	< 0.0001
7.6	Number of neurons in cerebral cortex	Number of neurons in rest of brain	Artiodactyls (n = 5)	$N_{CX} = e^{-14.59214} N_{ROB}^{1.904 \pm 0.172}$	0.976	0.0016
7.6	Number of neurons in cerebral cortex	Number of neurons in rest of brain	Afrotherians (n = 5), rodents (n = 10), eulipotyphlans (n = 5)	$N_{CX} = e^{-0.673368} N_{ROB}^{1.085 \pm 0.064}$	0.940	< 0.0001
8.1	Brain mass (grams)	Body mass (grams)	Eulipotyphlans (n = 5)	$M_{BR} = e^{-3.112349} M_{BD}^{0.727 \pm 0.094}$	0.952	0.0045
8.1	Brain mass (grams)	Body mass (grams)	Rodents (n = 10)	$M_{BR} = e^{-3.286723} M_{BD}^{0.712 \pm 0.071}$	0.927	< 0.0001
8.1	Brain mass (grams)	Body mass (grams)	Primates (n = 11, including humans)	$M_{BR} = e^{-3.323071} M_{BD}^{0.903 \pm 0.082}$	0.931	< 0.0001
8.1	Brain mass (grams)	Body mass (grams)	Afrotherians (n = 6)	$M_{BR} = e^{-2.92701} M_{BD}^{0.740 \pm 0.033}$	0.992	< 0.0001
8.1	Brain mass (grams)	Body mass (grams)	Artiodactyls (n = 4, excluding pig)	$M_{BR} = e^{-0.939437} M_{BD}^{0.548 \pm 0.038}$	0.990	0.0048

Appendix B: (continued)

Figure	Dependent variable	Independent variable	Species	Function	r^2	*p*-value
8.1	Brain mass (grams)	Body mass (grams)	Eulipotyphlans (n = 5), rodents (n = 10), primates (n = 11), scandentia (n = 1), afrotherians (n = 6), artiodactyls (n = 4, excluding pig)	$M_{BR} = e^{-3.114652} M_{BD}^{0.774 \pm 0.037}$	0.924	< 0.0001
8.2	Number of neurons in spinal cord	Body mass (grams)	Primates (n = 8)	$N_{SC} = e^{12.960178} M_{BD}^{0.359 \pm 0.028}$	0.965	< 0.0001
8.3	Number of neurons in facial motor nucleus	Body mass (grams)	Marsupials (n = 22)	$MN_{fac} = e^{7.4066354} M_{BD}^{0.184 \pm 0.015}$	0.884	< 0.0001
8.3	Number of neurons in facial motor nucleus	Body mass (grams)	Primates (n = 18)	$MN_{fac} = e^{7.7363026} M_{BD}^{0.127 \pm 0.033}$	0.484	0.0013
8.4	Number of neurons in rest of brain	Body mass (grams)	Primates (n = 12, including humans)	$N_{ROB} = e^{14.036812} M_{BD}^{0.525 \pm 0.089}$	0.777	0.0002
8.4	Number of neurons in rest of brain	Body mass (grams)	Afrotherians (n = 6), rodents (n = 10), eulipotyphlans (n = 5), artiodactyls (n = 5)	$N_{ROB} = e^{14.859394} M_{BD}^{0.321 \pm 0.022}$	0.898	< 0.0001
8.5	Number of neurons in cerebral cortex	Body mass (grams)	Primates (n = 12, including humans)	$N_{CX} = e^{14.19518} M_{BD}^{0.825 \pm 0.097}$	0.878	< 0.0001

Appendix B: (continued)

Figure	Dependent variable	Independent variable	Species	Function	r^2	*p*-value
8.5	Number of neurons in cerebral cortex	Body mass (grams)	Afrotherians (n = 6), rodents (n = 10), eulipotyphlans (n = 5), artiodactyls (n = 5)	$N_{CX} = e^{14.887552} M_{BD}^{0.462\pm0.020}$	0.954	< 0.0001
8.6	Number of neurons in cerebellum	Body mass (grams)	Primates (n = 13, including humans)	$N_{CB} = e^{15.656048} M_{BD}^{0.754\pm0.073}$	0.906	< 0.0001
8.6	Number of neurons in cerebellum	Body mass (grams)	Afrotherians (n = 5), rodents (n = 10), eulipotyphlans (n = 5), artiodactyls (n = 5)	$N_{CB} = e^{15.707664} M_{BD}^{0.561\pm0.034}$	0.919	< 0.0001
8.7	Neurons per milligram in rest of brain	Number of neurons in rest of brain	Afrotherians (n = 5), eulipotyphlans (n = 5), rodents (n = 10), artiodactyls (n = 5), scandentia (n = 1)	$DN_{ROB} = e^{25.128398} N_{ROB}^{-0.914\pm0.118}$	0.712	< 0.0001
8.8	Neurons per milligram in rest of brain	Body mass (grams)	Afrotherians (n = 6), eulipotyphlans (n = 5), rodents (n = 10), primates (n = 12, including humans), scandentia (n − 1), artiodactyls (n = 5)	$DN_{ROB} = e^{11.64696} M_{BD}^{-0.300\pm0.019}$	0.872	< 0.0001

Appendix B: (continued)

Figure	Dependent variable	Independent variable	Species	Function	r^2	*p*-value
8.9	Neurons per milligram in cerebral cortex	Body mass (grams)	Afrotherians (n = 6), eulipotyphlans (n = 5), rodents (n = 10), artiodactyls (n = 5), scandentia (n = 1)	$DN_{CX} = e^{11.986052} M_{BD}^{-0.294\pm0.014}$	0.944	< 0.0001
8.9	Neurons per milligram in cerebral cortex	Body mass (grams)	Primates (n = 12, including humans)	$DN_{CX} = e^{11.191639} M_{BD}^{-0.116\pm0.059}$	0.282	0.0757
8.10	Neurons per milligram in cerebellum	Body mass (grams)	Afrotherians (n = 5), rodents (n = 10), artiodactyls (n = 5),	$DN_{CB} = e^{13.805949} M_{BD}^{-0.156\pm0.017}$	0.808	< 0.0001
9.4	Mass of brain structure (grams)	Number of glial cells in brain structure	Afrotherians (n = 6), rodents (n = 10), eulipotyphlans (n = 5), artiodactyls (n = 5), primates (n = 12), scandentia (n = 1)	$M_{STR} = e^{-19.09168} O_{STR}^{1.054\pm0.016}$	0.974	< 0.0001
9.11	Glucose use per minute (micromoles)	Number of neurons in brain	Rodents (n = 3), primates (n = 3)	$Glu = e^{-18.72393} N_{BR}^{0.988\pm0.023}$	0.998	< 0.0001
9.12	Energy cost of brain (kilocalories per day)	Energy cost of body (kilocalories per day)	Primates (n = 11, including humans)	$E_{BR} = e^{-2.105168} M_{BD}^{1.036\pm0.122}$	0.889	< 0.0001
9.12	Energy cost of brain (kilocalories per day)	Energy cost of body (kilocalories per day)	Afrotherians (n = 5), rodents (n = 10), eulipotyphlans (n = 5)	$E_{BR} = e^{-1.828379} M_{BD}^{0.641\pm0.031}$	0.948	< 0.0001

Appendix B: (continued)

Figure	Dependent variable	Independent variable	Species	Function	r^2	p-value
10.1	Caloric intake (kilocalories per hour)	Body mass (kilograms)	Primates, nonhuman (n = 11)	$E_{IN} = e^{3.2328418} M_{BD}^{0.526 \pm 0.067}$	0.874	< 0.0001
10.2	Hours per day spent foraging and feeding	Body mass (kilograms)	Primates, nonhuman (n = 11)	$H = e^{1,015222} M_{BD}^{0.223 \pm 0.067}$	0.555	0.0085

Notes

Chapter 1

1. Bunnin and Yu, 2008, p. 289.
2. Edinger, 1908.
3. Kappers, Huber, and Crosby, 1936.
4. MacLean, 1964.
5. Sagan, 1977.
6. Carroll, 1988; Evans, 2000.
7. Shanahan et al., 2013.
8. Jenner, 2004.
9. Reviewed in Gould, 1977.
10. von Haller, 1762; Cuvier, 1801.
11. Snell, 1891.
12. Stephan and Andy, 1969.
13. Stephan and Andy, 1969.
14. Huxley, 1932.
15. Jerison, 1955.
16. Jerison, 1973.
17. Marino, 1998; Sol et al., 2005.
18. Marino, 1998.
19. Deaner et al., 2007.
20. Marino, 1998; Herculano-Houzel, 2011.

21. Marino, 1998.

22. Roth and Dicke, 2005.

23. Deaner et al., 2007.

24. MacLean et al., 2014.

25. Mink, Blumenschine, and Adams, 1981.

26. Nimchinsky et al., 1999.

27. Williams, 2012.

28. Evrard, Forro, and Logothetis, 2012.

29. Butti et al., 2009.

30. Oberheim et al., 2009.

31. Han et al., 2013.

32. Spocter et al., 2012.

33. Mantini et al., 2013.

34. Sallet et al., 2013.

35. Shanahan et al., 2013.

36. Evans et al., 2004; Dumas et al., 2012.

37. Dennis et al., 2012; Charrier et al., 2012.

38. Enard et al., 2009.

39. Somel, Xiling, and Khaitovich, 2013.

40. Prabhakar et al., 2008.

Chapter 2

1. Tower, 1954; Tower and Elliott, 1952.

2. Haug, 1987.

3. Williams and Herrup, 1988.

4. Elias and Schwartz, 1971; Stephan, Frahm, and Baron, 1981; Hofman, 1985.

5. Tower and Elliott, 1952; Tower, 1954; Haug, 1987; Stolzenburg, Reichenbach, and Neumann, 1989.

6. Kandel, Schwartz, and Jessel, 2000, p. 20.

7. Hawkins and Olszewski, 1957; Andersen, Korbo, and Pakkenberg, 1992.

8. Herculano-Houzel, 2002.

9. Lee, Thornthwait, and Rasch, 1984.

10. Mullen, Buck, and Smith, 1992.

11. Herculano-Houzel and Lent, 2005.

12. Bahney and von Bartheld, 2014; Miller et al., 2014.

13. Collins et al., 2010; Young et al., 2012.

14. Bahney and von Bartheld, 2014; Miller et al., 2014.

15. Herculano-Houzel et al., 2015a.

Chapter 3

1. Collins et al., 2013; Wong et al., 2013.

2. Burish et al., 2010.

3. Azevedo et al., 2009.

4. Tower and Elliott, 1952; Tower, 1954.

Chapter 4

1. Herculano-Houzel, Mota, and Lent, 2006.

2. Herculano-Houzel et al., 2007.

3. Herculano-Houzel et al., 2007.

4. Herculano-Houzel, Mota, and Lent, 2006.

5. Herculano-Houzel et al., 2011; Gabi et al., 2010.

6. Herculano-Houzel, 2010.

7. Herculano-Houzel, 2012.

8. Murphy et al., 2001, 2004.

9. Alvarez et al., 1980.

10. Murphy et al., 2001, 2004.

11. Douady et al., 2002.

12. Herculano-Houzel, Manger, and Kaas, 2014a.

13. Rowe, Macrini, and Luo, 2011.

14. Bloch, Rose, and Gingerich, 1998.

15. Luo, Compton, and Sun, 2001.

16. Herculano, Kaas and Manger, 2014.

17. Silcox, Dalmyn, and Bloch, 2009.

18. Silcox, Dalmyn, and Bloch, 2009.

19. Gabi et al., 2010.

20. Mota and Herculano-Houzel, 2012.

21. Mota and Herculano-Houzel, 2012; Ventura-Antunes and Herculano-Houzel, 2013.

22. Lange, 1975; Jacobs et al., 2014.

Chapter 5

1. Herculano-Houzel, Mota, and Lent, 2006.

2. Azevedo et al., 2009.

3. Herculano-Houzel and Kaas, 2011.

4. De Sousa and Wood, 2007.

Chapter 6

1. Twain, 1973.

2. Seehausen, 2002.

3. Iriki, Tanaka, and Iwamura, 1996.

4. Weir, Chappell, and Kacelnik, 2002; Auersperg et al., 2012; Klump et al., 2015.

5. Pepperberg, 1999.

6. Wise, 2003; Johnson, 1993.

7. Inoue and Matsuzawa, 2007.

8. Plotnik et al., 2011; Brosnan and de Waal, 2002.

9. Byrne and Corp, 2004; Kirkpatrick, 2007.

10. Emery and Clayton, 2001.

11. Prior, Schwartz, and Güntürkün, 2008.

12. Deaner et al., 2007.
13. MacLean et al., 2014.
14. E.g., Ramnani, 2006.
15. Esteves, 2013.
16. Herculano-Houzel et al., 2014.
17. Herculano-Houzel and Kaas, 2011.
18. Maseko et al., 2012.
19. Cunha et al., submitted.
20. Marino and Frohoff, 2011.
21. Manger, 2013.
22. Reiss and Marino, 2001.
23. Yaman et al., 2012.
24. King and Janik, 2013.
25. Bruck, 2013.
26. Eriksen and Pakkenberg, 2007.
27. Walloe et al., 2010.
28. Schmitz and Hof, 2000.

Chapter 7

1. Hofman, 1985; Stephan, Frahm, and Baron, 1981; Rilling and Insel, 1999.
2. Clark, Mitra, and Wang, 2001.
3. Herculano-Houzel et al., 2014.
4. Passingham, 2012.
5. Brodmann, 1912.
6. Semendeferi, et al., 2002.
7. Schoenemann, Sheehan and Glotzer, 2005.
8. Smaers et al., 2011.
9. Barton and Venditti, 2013.
10. Ribeiro et al., 2013.
11. Gabi et al., submitted.

12. Herculano-Houzel, Watson, and Paxinos, 2013.

13. Shanahan et al., 2013.

14. Cragg, 1967; Collonier and O'Kusky, 1981; Schüz and Palm, 1989; Schüz and Demianenko, 1995; Braitenberg and Schüz, 1991.

15. See Alex Wissner-Gross's video "A New Equation for Intelligence," at https://www.ted.com/talks/alex_wissner_gross_a_new_equation_for_intelligence?

16. Morgane, Jacobs, and MacFarland, 1980.

Chapter 8

1. Calder, 1996.

2. von Haller, 1762; Snell, 1891.

3. Kleiber, 1932, 1947.

4. Jerison, 1973; Martin, 1996.

5. Burish et al., 2010.

6. Fu et al., 2012.

7. Herculano-Houzel, Kaas, and de Oliveira-Souza, 2015.

8. Burish et al., 2010.

9. Watson, Provis, and Herculano-Houzel, 2012; Sherwood, 2005.

10. Watson, Provis, and Herculano-Houzel, 2012.

11. E.g., Hollyday and Hamburger, 1976.

12. Tanaka and Landmesser, 1986.

13. Burish et al., 2010.

14. Herculano-Houzel, 2015.

15. Herculano-Houzel and Kaas, 2011.

16. Lloyd, 2013.

17. Mota and Herculano-Houzel, 2014.

Chapter 9

1. Kety, 1957; Sokoloff, 1960; Rolfe and Brown, 1997; Clarke and Sokoloff, 1999.

2. Mink, Blumenschine, and Adams, 1981.

3. Pellerin and Magistretti, 2004.

4. Aiello and Wheeler, 2006.

5. Hofman, 1983.

6. Mink, Blumenschine, and Adams, 1981.

7. Herculano-Houzel, 2011.

8. Cáceres et al., 2003; Uddin et al., 2004.

9. Kleiber, 1932, 1947.

10. Karbowski, 2007; Attwell and Laughlin, 2001.

11. Hawkins and Olzewski, 1957.

12. Kast, 2001.

13. Zimmer, 2009.

14. Kandel, Schwartz, and Jessel, 2000, p. 20.

15. Bear, Connors, and Paradiso, 2006.

16. E.g., Nedergaard, Ransom, and Goldman, 2003; Allen and Barres, 2009.

17. Reviewed in Allen and Barres, 2009.

18. Magistretti, 2006; Lee et al., 2012.

19. Nissl, 1898.

20. Friede, 1954.

21. Tower and Elliott, 1952.

22. Hawkins and Olzewski, 1957.

23. Haug, 1987.

24. Tower, 1954.

25. von Bartheld et al., submitted.

26. Azevedo et al., 2009.

27. Herculano-Houzel, 2014.

28. Olszewski et al., submitted.

29. Bandeira, Lent and Herculano-Houzel, 2009.

30. Mota and Herculano-Houzel, 2014.

31. Magistretti, 2006.

32. Attwell and Laughlin, 2001.

33. Karbowski, 2007.

34. Herculano-Houzel, 2011.

35. Porter and Brand, 1995a, 1995b.

36. Cáceres et al., 2003; Uddin et al., 2004.

37. Smith et al., 2002.

38. Fox and Raichle, 1986.

39. Lin et al., 2010.

40. Shulman, Hyder, and Rothman, 2009.

41. La Fougère et al., 2009; d'Avila et al., 2008; Finsterer, 2008; Zhao et al., 2008.

42. Wilson et al., 2007.

43. Lennie, 2003; Kerr, Greenberg, and Helmchen, 2005; Shoham, O'Connor, and Segev, 2006).

44. Gilestro, Tononi, and Cirelli, 2009; Turrigiano, 2008.

Chapter 10

1. Fonseca-Azevedo and Herculano-Houzel, 2012.

2. Owen-Smith, 1988.

3. Knott, 1998.

4. Herculano-Houzel, 2011.

5. Knott, 1998.

6. Watts, 1988.

7. Herculano-Houzel, 2011.

Chapter 11

1. Fonseca-Azevedo and Herculano-Houzel, 2012; Organ et al., 2011.

2. Organ et al., 2011.

3. Lieberman, 2013.

4. Bramble and Liebermann, 2004.

5. Berna et al., 2012; Gowlett et al., 1981.

6. Carmody, Weintraub, and Wrangham, 2011.

7. Susman, 1998; Tocheri et al., 2008; Alba, Moyà-Solà, and Köhler, 2003.

8. Wrangham, 2009.

9. Fonseca-Azevedo and Herculano-Houzel, 2012.

10. Wobber, Hare, and Wrangham, 2008.

11. Warneken and Rosati, 2015.

12. Diamond, 1987.

13. Fonseca-Azevedo and Herculano-Houzel, 2012.

14. Adler, 2013, p. 44.

Chapter 12

1. Krutzen et al., 2014; Cantor and Whitehead, 2013.

2. Whiten et al., 1999.

3. Raji et al., 2010.

4. Brown et al., 2012; Yellen et al., 1995; Wadley, Hodgkins, and Grant, 2009; Lieberman, 2013.

5. Pearce, Stringer, and Dunbar, 2013.

6. Lieberman, 2013.

7. Krause et al., 2007; Coop et al., 2008.

8. Higham et al., 2014.

9. Prüfer et al., 2014.

10. Standage, 2009.

Epilogue

1. Azevedo et al., 2009.

References

Adams D. 2005. *The restaurant at the end of the universe.* New York, Ballantine Books.

Adler J. 2013. The mind on fire. *Smithsonian* (June): 43–45.

Aiello LC, Wheeler P. 2006. The expensive-tissue hypothesis. *Curr Anthropol* 36: 199–221.

Alba D, Moyà-Solà S, Köhler S. 2003. Morphological affinities of the *Australopithecus afarensis* hand on the basis of manual proportions and relative thumb length. *J Hum Evol* 44: 225–254.

Allen NJ, Barres BA. 2009. Glia: More than just brain glue. *Nature* 457: 675–677.

Alvarez LW, Alvarez W, Asaro F, Michel HV. 1980. Extraterrestrial cause for the Cretaceous-Tertiary extinction. *Science* 208: 1095–1108.

Andersen BB, Korbo L, Pakkenberg B. 1992. A quantitative study of the human cerebellum with unbiased stereological techniques. *J Comp Neurol* 326: 549–560.

Armstrong E. 1990. Brains, bodies and metabolism. *Brain Behav Evol* 36: 166–176.

Aschoff J, Günther B, Kramer K. 1971. *Energiehaushalt und Temperaturregulation.* Munich, Urban and Schwarzenberg.

Attwell D, Laughlin SB. 2001. An energy budget for signaling in the grey matter of the brain. *J Cereb Blood Flow Metab* 21: 1133–1145.

Auersperg AM, Szabo B, von Bayern AM, Kacelnik A. 2012. Spontaneous innovation in tool manufacture and use in a Goffin's cockatoo. *Curr Biol* 22: R903–R904.

Azevedo FAC, Carvalho LRB, Grinberg LT, Farfel JM, Ferretti REL, Leite REP, et al. 2009. Equal numbers of neuronal and non-neuronal cells make the human brain an isometrically scaled-up primate brain. *J Comp Neurol* 513: 532–541.

Bahney J, von Bartheld CS. 2014. Validation of the isotropic fractionator: Comparison with unbiased stereology and DNA extraction for quantification of glial cells. *J Neurosci Methods* 222: 165–174.

Barton RA, Harvey PH. 2000. Mosaic evolution of brain structure in mammals. *Nature* 405: 1055–1058.

Barton RA, Venditti C. 2013. Human frontal lobes not relatively large. *Proc Natl Acad Sci USA* 110: 9001–9006.

Bear MF, Connors B, Paradiso M. 2006. *Neuroscience: Exploring the brain.* 3rd edition. Philadelphia: Lippincott Williams & Wilkins.

Berna F, Goldberg P, Horwitz LK, Brink J, Holt S, Bamford M, Chazan M. 2012. Microstratigraphic evidence of in situ fire in the Acheulean strata of Wonderwerk Cave, Northern Cape province, South Africa. *Proc Natl Acad Sci USA* 109: E1215–E1220.

Bloch JI, Rose KD, Gingerich PD. 1998. New species of *Batonoides* (Lipotyphla, Geolabididae) from the early Eocene of Wyoming: Smallest known mammal? *J Mammal* 79: 804–827.

Braitenberg V, Schüz A. 1991. *Cortex: Statistics and geometry of neuronal connectivity.* Berlin, Springer.

Bramble DM, Lieberman DE. 2004. Endurance running and the evolution of *Homo. Nature* 432: 345–352.

Brodmann K. 1912. *Localisation in the cerebral cortex.* Berlin,Springer.

Brosnan SF, de Waal FBM. 2002. A proximate perspective on reciprocal altruism. *Hum Nat* 13: 129–152.

Brown KS, Mareau CW, Jacobs Z, Schoville BJ, Oestmo S, Fisher EC, et al. 2012. An early and enduring advanced technology originating 71,000 years ago in South Africa. *Nature* 491: 590–593.

Bunnin N, Yu J. 2008. *Blackwell dictionary of Western philosophy.* New York, Wiley-Blackwell.

Burish MJ, Peebles JK, Tavares L, Baldwin M, Kaas JH, Herculano-Houzel S. 2010. Cellular scaling rules for primate spinal cords. *Brain Behav Evol* 76: 45–59.

Butti C, Sherwood CC, Hakeem AY, Allman JM, Hof PR. 2009. Total number and volume of von Economo neurons in the cerebral cortex of cetaceans. *J Comp Neurol* 515: 243–259.

Bruck JN. 2013. Decades-long social memory in bottlenose dolphins. *Proc Royal Soc B Biol Sci* 280: 20131.

Byrne R, Corp N. 2004. Neocortex size predicts deception rate in primates. *Proc Biol Sci* 271: 1693–1699.

Cáceres M, Lachuer J, Zapala MA, Redmond JC, Kudo L, Geschwind DH, et al. 2003. Elevated gene expression levels distinguish human from non-human primate brains. *Proc Natl Acad Sci USA* 100: 13030–13035.

Calder WA. 1996. *Size, function and life history*. New York, Dover.

Cantor M, Whitehead H. 2013. The interplay between social networks and culture: Theoretically and among whales and dolphins. *Phil Trans Roy Soc B Biol Sci* 368: doi:10.1098/rstb.2012.0340.

Carmody RN, Weintraub GS, Wrangham RW. 2011. Energetic consequences of thermal and nonthermal food processing. *Proc Natl Acad Sci USA* 108: 19199–19203.

Carroll RL. 1988. *Vertebrate paleontology and evolution*. New York, W.H. Freeman & Company.

Charrier C, Joshi K, Coutinho-Budd J, Kim JE, Lambert N, de Marchena J, et al. 2012. Inhibition of SRGAP2 function by its human-specific paralogs induces neoteny during spine maturation. *Cell* 149: 923–935.

Clark DA, Mitra PP, Wang SS. 2001. Scalable architecture in mammalian brains. *Nature* 411: 189–193.

Clarke DD, Sokoloff L. 1999. Circulation and energy metabolism of the brain. In *Basic neurochemistry: Molecular, cellular and medical aspects*. Edited by Siegel GJ, Agranoff BW, Albers RW, Fisher SK, Uhler MD. Philadelphia, Lippincott-Raven, pp. 637–669.

Collins CE, Leitch DB, Wong P, Kaas JH, Herculano-Houzel S. 2013. Faster scaling of visual neurons in cortical areas relative to subcortical structures in primate brains. *Brain Struct Funct* 218: 805–816.

Collins CE, Young NA, Flaherty DK, Airey DC, Kaas JH. 2010. A rapid and reliable method of counting neurons and other cells in brain tissue: A comparison of flow cytometry and manual counting methods. *Front Neuroanat* 4: 5.

Colonnier M, O'Kusky J. 1981. Number of neurons and synapses in the visual cortex of different species. *Rev Can Biol* 40: 91–99.

Coop G, Bullaughey K, Luca F, Przeworski M. 2008. The timing of selection at the human FOXP2 gene. *Mol Biol Evol* 25: 1257–1259.

Cragg BG. 1967. The density of synapses and neurones in the motor and visual areas of the cerebral cortex. *J Anat* 101: 639–654.

Cunha FB, Pettigrew J, Manger PR, Herculano-Houzel S. Submitted. Echolocating Microchiroptera have a smaller cerebral cortex, not a larger cerebellum, compared to Macrochiroptera.

Cuvier G. 1801. *Leçons d'anatomie comparée*. Paris, Baudouin.

Darwin C. 1859. *On the origin of species by means of natural selection, or, The preservation of favoured races in the struggle for life*. London, John Murray.

d'Avila JC, Santiago AP, Amâncio RT, Galina A, Oliveira MF, Bozza FA. 2008. Sepsis induces brain mitochondrial dysfunction. *Crit Care Med* 36: 1925–1932.

Deaner RO, Isler K, Burkart J, van Schaik C. 2007. Overall brain size, not encephalization quotient, best predicts cognitive ability across non-human primates. *Brain Behav Evol* 70: 115–124.

Deaner RO, van Schaik CP, Johnson V. 2006. Do some taxa have better domain-general cognition than others? A meta-analysis of non-human primate studies. *Evol Psychol* 4: 149–196.

Dennis MY, Nuttle X, Sudmant PH, Antonacci F, Graves TA, Nefedov M, et al. 2012. Evolution of human-specific neural SRGAP2 genes by incomplete segmental duplication. *Cell* 149: 912–922.

De Sousa A, Woods B. 2007. The hominin fossil record and the emergence of the human central nervous system. In *Evolution of nervous systems: A comprehensive reference*, volume 4. Edited by Kaas JH, Preuss TM Oxford, Elsevier, pp. 291–336.

Diamond J. 1987. The worst mistake in the history of the human race. *Discover*, May: 64–66.

Douady CJ, Chatelier CI, Madsen O, de Jong WW, Catzeflis F, Springer MS, Stanhope MJ. 2002. Molecular phylogenetic evidence confirming the Eulipotyphla concept and in support of hedgehogs as the sister group to shrews. *Mol Phylogenet Evol* 25: 200–209.

Dubois E. 1897. Sur le rapport du poids de l'encéphale avec la grandeur du corps chez mammifères. *Bull Soc Anthropol Paris* 8: 337–376.

Dumas L, O'Bleness MS, Davis JM, Dickens CM, Anderson N, Keeney JG, et al. 2012. DUF1220-domain copy number implicated in human brain-size pathology and evolution. *Am J Hum Genet* 91: 444–454.

Edinger L. 1908. The relations of comparative anatomy to comparative psychology. *Comp Neurol Psychol* 18: 437–457.

Elias H, Schwartz D. 1971. Cerebro-cortical surface areas, volumes, lengths of gyri and their interdependence in mammals, including man. *Z Saugetierkd* 36: 147–163.

Emery NJ, Clayton NS. 2001. Effects of experience and social context on prospective caching strategies by scrub jays. *Nature* 414: 443–446.

Enard W, Gehre S, Hammerschmidt K, Hölter SM, Blass T, Somel M, et al. 2009. A humanized version of Foxp2 affects cortico-basal ganglia circuits in mice. *Cell* 137: 961–971.

Eriksen N, Pakkenberg B. 2007. Total neocortical cell number in the mysticete brain. *Anat Rec* 290: 83–95.

Esteves B. 2013. O cru, o cozido e o cérebro. *Revista Piauí* 71: 64–69.

Evans PD, Anderson JR, Vallender EJ, Choi SS, Lahn BT. 2004. Reconstructing the evolutionary history of microcephalin, a gene controlling brain size. *Hum Mol Genet* 13: 1139–1145.

Evans SE. 2000. General discussion: Amniote evolution. In *Evolutionary developmental biology of the cerebral cortex*. Edited by Bock GR, Cardew G. John Wiley & Sons, Chichester, pp. 109–113.

Evrard HC, Forro T, Logothetis NK. 2012. Von Economo neurons in the anterior insula of the macaque monkey. *Neuron* 74: 482–489.

Finlay BL, Darlington RB. 1995. Linked regularities in the development and evolution of mammalian brains. *Science* 268: 1578–1584.

Finsterer J. 2008. Cognitive decline as a manifestation of mitochondrial disorders (mitochondrial dementia). *J Neurol Sci* 272: 20–33.

Fonseca-Azevedo K, Herculano-Houzel S. 2012. Metabolic constraint imposes tradeoff between body size and number of brain neurons in human evolution. *Proc Natl Acad Sci USA* 109: 18571–18576.

Fox PT, Raichle ME. 1986. Focal physiological uncoupling of cerebral blood flow and oxidative metabolism during somatosensory stimulation in human subjects. *Proc Natl Acad Sci USA* 83: 1140–1144.

Frahm HD, Stephan H, Stephan M. 1982. Comparison of brain structure volumes in Insectivora and Primates. 1. Neocortex. *J Hirnforsch* 23: 375–389.

Friede R. 1954. Der quantitative Anteil der Glia an der Cortexentwicklung. *Acta Anat (Basel)* 20: 290–296.

Fu Y, Rusznák Z, Herculano-Houzel S, Watson C, Paxinos G. 2012. The mouse central nervous system: Age-dependent changes in cellular composition characterizing postnatal development and maturation. *Brain Struct Funct* 218: 1337–1354.

Gabi M, Collins CE, Wong P, Torres LB, Kaas JH, Herculano-Houzel S. 2010. Cellular scaling rules for the brains of an extended number of primate species. *Brain Behav Evol* 76: 32–44.

Gabi M, Neves K, Masseron C, Ventura-Antunes L, Ribeiro P, Torres L, et al. Submitted. No expansion in numbers of prefrontal neurons in primate and human evolution.

Gilestro GF, Tononi G, Cirelli C. 2009. Widespread changes in synaptic markers as a function of sleep and wakefulness in Drosophila. *Science* 324: 109–112.

Gombrich EH. 2005. *A little history of the world.* Translated by Mustill C. New Haven, Yale University Press.

Goodall H. 2014. *The story of music: From Babylon to the Beatles; How music shaped civilization.* New York, Pegasus.

Gould SJ. 1977. *Ontogeny and phylogeny.* Cambridge, Massachusetts, Harvard University Press.

Gowlett JAJ, Harris JWK, Walton D, Wood BA.1981. Early archaeological sites, hominid remains, and traces of fire from Chesowanja, Kenya. *Nature* 294: 125–129.

Grossman LI, Schmidt TR, Wildman DE, Goodman M. 2001. Molecular evolution of aerobic energy metabolism in primates. *Mol Phylogenet Evol* 18: 26–36.

Güntürkün O. 2014. Is dolphin cognition special? *Brain Behav Evol* 83: 177–180.

Haeckel E. 1886. *Natürliche Schöpfungsgeschichte.* Berlin, Georg Reimer.

Han X, Chen M, Wang F, Windrem M, Wang S, Shanz S, et al. 2013. Forebrain engraftment by human glial progenitor cells enhances synaptic plasticity and learning in adult mice. *Stem Cells* 12: 342–353.

Haug H. 1987. Brain sizes, surfaces, and neuronal sizes of the cortex cerebri: A stereological investigation of man and his variability and a comparison with some mammals (primates, whales, marsupials, insectivores, and one elephant). *Am J Anat* 180: 126–142.

Hawkins A, Olszewski J. 1957. Glia/nerve cell index for cortex of the whale. *Science* 126: 76–77.

Herculano-Houzel S. 2002. Do you know your brain? A survey on public neuroscience literacy at the closing of the decade of the brain. *Neuroscientist* 8: 98–110.

Herculano-Houzel S. 2010. Coordinated scaling of cortical and cerebellar numbers of neurons. *Front Neuroanat* 4: 12.

Herculano-Houzel S. 2011. Not all brains are made the same: New views on brain scaling in evolution. *Brain Behav Evol* 78: 22–36.

Herculano-Houzel S. 2012. Neuronal scaling rules for primate brains: The primate advantage. *Prog Brain Res* 195: 325–340.

Herculano-Houzel S. 2014. The glia/neuron ratio: How it varies uniformly across brain structures and species and what that means for brain physiology and evolution. *Glia* 62: 1377–1391.

Herculano-Houzel S. 2015. Decreasing sleep requirement with increasing numbers of neurons as a driver for bigger brains and bodies in mammalian evolution. *Proc Biol Sci*. In press.

Herculano-Houzel S, Avelino-de-Souza K, Neves K, Porfírio J, Messeder D, Calazans I,et al. 2014. The elephant brain in numbers. *Front Neuroanat* 8: 46.

Herculano-Houzel S, Collins CE, Wong P, Kaas JH. 2007. Cellular scaling rules for primate brains. *Proc Natl Acad Sci USA* 104: 3562–3567.

Herculano-Houzel S, Kaas JH. 2011. Gorilla and orangutan brains conform to the primate scaling rules: Implications for human evolution. *Brain Behav Evol* 77: 33–44.

Herculano-Houzel S, Kaas JH, de Oliveira-Souza R. 2015. Corticalization of motor control in humans is a consequence of brain scaling in primate evolution. *J Comp Neurol* 523: doi:10.1002/cne.23792.

Herculano-Houzel S, Kaas JH, Miller D, Von Bartheld CS. 2015a. How to count cells: The advantages and disadvantages of the isotropic fractionator compared with stereology. *Cell Tissue Res* 360: 19–42.

Herculano-Houzel S, Lent R. 2005. Isotropic fractionator: A simple, rapid method for the quantification of total cell and neuron numbers in the brain. *J Neurosci* 25: 2518–2521.

Herculano-Houzel S, Manger PR, Kaas JH. 2014. Brain scaling in mammalian brain evolution as a consequence of concerted and mosaic changes in numbers of neurons and average neuronal cell size. *Front Neuroanat* 8: 77.

Herculano-Houzel S, Messeder D, Fonseca-Azevedo K, Araujo Pantoja N. 2015b. When larger brains do not have more neurons: Intraspecific increase in numbers of cells is compensated by decreased average cell size. *Front Neuroanat* 9: 64.

Herculano-Houzel S, Mota B, Lent R. 2006. Cellular scaling rules for rodent brains. *Proc Natl Acad Sci USA* 103: 12138–12143.

Herculano-Houzel S, Ribeiro PFM, Campos L, da Silva AV, Torres LB, Catania KC, Kaas JH. 2011. Updated neuronal scaling rules for the brains of Glires (rodents/lagomorphs). *Brain Behav Evol* 78: 302–314.

Herculano-Houzel S, Watson C, Paxinos G. 2013. Distribution of neurons in functional areas of the mouse cerebral cortex reveals quantitatively different cortical zones. *Front Neuroanat* 7: 35.

Higham T, Douka K, Wood R, Ramsey CB, Brock F, Bassell L, et al. 2014. The timing and spatiotemporal patterning of Neanderthal disappearance. *Nature* 512: 306–309.

Hofman MA. 1983. Energy metabolism, brain size and longevity in mammals. *Q Rev Biol* 58: 495–512.

Hofman MA. 1985. Size and shape of the cerebral cortex in mammals. I. The cortical surface. *Brain Behav Evol* 27: 28–40.

Hollyday M, Hamburger V. 1976. Reduction of the naturally occurring motor neuron loss by enlargement of the periphery. *J Comp Neurol* 170: 311–320.

Huxley JS. 1932. *Problems of relative growth.* London, Allen & Unwin.

Inoue S, Matsuzawa T. 2007. Working memory of numerals in chimpanzees. *Curr Biol* 17: R1004–R1005.

Iriki A, Tanaka M, Iwamura Y. 1996. Coding of modified body schema during tool use by macaque postcentral neurons. *Neuroreport* 7: 2325–2330.

Jacobs B, Johnson NL, Wahl D, Schall M, Maseko BC, Lewandowski A, et al. 2014. Comparative neuronal morphology of the cerebellar cortex in afrotherians, carnivores, cetartiodactyls, and primates. *Front Neuroanat* 8: 24.

Jenner RA. 2004. When molecules and morphology clash: Reconciling conflicting phylogenies of the Metazoa by considering secondary character loss. *Evol Dev* 6: 372–378.

Jerison HJ. 1955. Brain to body ratios and the evolution of intelligence. *Science* 121: 447–449.

Jerison HJ. 1973. *Evolution of the brain and intelligence.* New York, Academic Press.

Johnson LE. 1993. *A morally deep world: An essay on moral significance and environmental ethics.* Cambridge: Cambridge University Press.

Kandel ER, Schwartz JH, Jessel TM. 2000. *Principles of neural science.* 4th edition. New York, McGraw-Hill.

Kappers CA, Huber CG, Crosby EC. 1936. *Comparative anatomy of the nervous system of vertebrates, including man.* New York, Hafner.

Karbowski J. 2007. Global and regional brain metabolic scaling and its functional consequences. *BMC Biol* 5: 18.

Kast B. 2001. The best supporting actors. *Nature* 412: 674–676.

Kerr JN, Greenberg D, Helmchen F. 2005. Imaging input and output of neocortical networks in vivo. *Proc Natl Acad Sci USA* 102: 14063–14068.

Kety SS. 1957. The general metabolism of the brain in vivo. In *Metabolism of the nervous system.* Edited by Richter D. London, Pergamon, pp. 221–237.

King SL, Janik VM. 2013. Bottlenose dolphins can use learned vocal labels to address each other. *Proc Natl Acad Sci USA* 110: 13216–13221.

Kirkpatrick C. 2007. Tactical deception and the great apes: Insight into the question of theory of mind. *Totem U West Ontario J Anthropol* 15: 4.

Kleiber M. 1932. Body size and metabolism. *Hilgardia* 6: 315–353.

Kleiber M. 1947. Body size and metabolic rate. *Physiol Rev* 27: 511–541.

Klump BC, van der Wal JEM, St Clair JJH, Rutz C. 2015. Context-dependent "safe-keeping" of foraging tools in New Caledonian crows. *Proc Biol Sci* 282: 20150278.

Knott CD. 1998. Changes in orangutan caloric intake, energy balance, and ketones in response to fluctuating fruit availability. *Int J Primatol* 19: 1061–1079.

Krause J, Lalueza-Fox C, Orlando L, Enard W, Green RE, Burbano HA, et al. 2007. The derived FOXP2 variant of modern humans was shared with Neanderthals. *Curr Biol* 17: 1908–1912.

Krutzen M, Kreicker S, Macleod CD, Learmonth J, Kopps AM, Walsham P, Allen SJ. 2014. Cultural transmission of tool used by Indo-Pacific bottlenose dolphins (*Tursiops sp.*) provides access to a novel foraging niche. *Proc Royal Soc B Biol Sci* 281: doi:10.1098/rspb.2014.0374.

la Fougère C, Rominger A, Förster S, Geisler J, Bartenstein P. 2009. PET and SPECT in epilepsy: A critical review. *Epilepsy Behav* 15: 50–55.

Lange W. 1975. Cell number and cell density in the cerebellar cortex of man and some other mammals. *Cell Tissue Res* 157: 115–124.

Lee GM, Thornthwait JT, Rasch EM. 1984. Picogram per cell determination of DNA by flow cytofluorometry. *Anal Biochem* 137: 221–226.

Lee Y, Morrison BM, Li Y, Lengacher S, Farah MH, Hoffman PN, et al. 2012. Oligodendroglia metabolically support axons and contribute to neurodegeneration. *Nature* 487: 443–448.

Lennie P. 2003. The cost of cortical computation. *Curr Biol* 13: 493–497.

Lieberman DE. 2013. *The story of the human body.* New York, Vintage Books.

Lin A-L, Fox PT, Hardies J, Duong TQ, Gao J-H. 2010. Nonlinear coupling between cerebral blood flow, oxygen consumption, and ATP production in human visual cortex. *Proc Natl Acad Sci USA* 107: 8446–8451.

Lloyd AC. 2013. The regulation of cell size. *Cell* 154: 1194–1205.

Luo Z-X, Crompton AW, Sun AL. 2001. A new mammaliaform form the early Jurassic and evolution of mammalian characteristics. *Science* 292: 1535–1540.

MacLean EL, Hare B, Nunn CL, Adessi E, Amici F, Anderson RC, et al. 2014. The evolution of self-control. *Proc Natl Acad Sci USA* 111: E2140–E2148.

MacLean PD. 1964. Man and his animal brains. *Mod Med* 2: 95–106.

MacLean PD. 1990. *The triune brain in evolution: Role in paleocerebral functions.* Plenum Press, New York.

Magistretti PJ. 2006. Neuron-glia metabolic coupling and plasticity. *J Exp Biol* 209: 2304–2311.

Mandel ESJ. 2014. *Station eleven: A novel.* New York, Knopf.

Manger PR. 2013. Questioning the interpretations of behavioural observations of cetaceans: Is there really support for a special intellectual status for this mammalian order? *Neurosci* 250: 664–696.

Mantini D, Corbetta M, Romani GL, Orban GA, Vanduffel W. 2013. Evolutionarily novel functional networks in the human brain? *J Neurosci* 33: 3259–3275.

Marino L. 1998. A comparison of encephalization between odontocete cetaceans and anthropoid primates. *Brain Behav Evol* 51: 230–238.

Marino L, Frohoff T. 2011. Towards a new paradigm of non-captive research on cetacean cognition. *PLoS One* 6: e24121.

Martin RD. 1996. Scaling of the mammalian brain: The maternal energy hypothesis. *News Physiol Sci* 11: 149–156.

Maseko BC, Spocter MA, Haagensen M, Manger PR. 2012. Elephants have relatively the largest cerebellum size of mammals. *Anat Rec* 295: 661–672.

Miller DJ, Balaram P, Young NA, Kaas JH. 2014. Three counting methods agree on cell and neuron number in chimpanzee primary visual cortex. *Front Neuroanat* 8: 36.

Mink JW, Blumenschine RJ, Adams DB. 1981. Ratio of central nervous system to body metabolism in vertebrates: Its constancy and functional basis. *Am J Physiol* 241: 203–212.

Miodownik M. 2014. *Stuff matters: Exploring the marvelous materials that shape our man-made world.* Boston, Houghton Mifflin Harcourt.

Morgane PJ, Jacobs MS, MacFarland WL. 1980. The anatomy of the brain of the bottlenose dolphin (*Tursiops truncatus*). Surface configurations of the telencephalon of the bottlenose dolphin with comparative anatomical observations in four other cetacean species. *Brain Res Bull* 5(suppl): 1–107.

Mota B, Herculano-Houzel S. 2012. How the cortex gets its folds: An inside-out, connectivity-driven model for the scaling of mammalian cortical folding. *Front Neuroanat* 6: 3.

Mota B, Herculano-Houzel S. 2014. All brains are made of this: A fundamental building block of brain matter with matching neuronal and glial masses. *Front Neuroanat* 8: 127.

Mullen RJ, Buck CR, Smith AM. 1992. NeuN, a neuronal specific nuclear protein in vertebrates. *Development* 116: 201–211.

Murphy WJ, Eizirik E, Johnson WE, Ping Zhang Y, Ryder OA, O'Brien SJ. 2001. Molecular phylogenetics and the origins of placental mammals. *Nature* 409: 614–618.

Murphy WJ, Pevzner PA, O'Brien SJ. 2004. Mammalian phylogenomics comes of age. *Trends Genet* 20: 631–639.

Nedergaard M, Ransom B, Goldman SA. 2003. New roles for astrocytes: Redefining the functional architecture of the brain. *Trends Neurosci* 26: 523–530.

Neves K, Ferreira Meireles F, Tovar-Moll F, Gravett N, Bennett NC, Kaswera C, Gilissen E, Manger PR, Herculano-Houzel S. 2014. Cellular scaling rules for the brain of afrotherians. *Front Neuroanat* 8: 5.

Nimchinsky EA, Gilissen E, Allman JM, Perl DP, Erwin JM, Hof PR. 1999. A neuronal morphologic type unique to humans and great apes. *Proc Natl Acad Sci USA* 96: 5268–5273.

Nissl F. 1898. Nervenzellen und graue Substanz. *Münch med Wochenschr* 45: 988–992, 1023–1029, 1060–1063.

Oberheim NA, Takano T, Han X, He W, Lin JHC, Wang F, et al. 2009. Uniquely hominid features of adult human astrocytes. *J Neurosci* 29: 3276–3287.

Organ C, Nunn CL, Machanda Z, Wrangham RW. 2011. Phylogenetic rate shifts in feeding time during the evolution of Homo. *Proc Natl Acad Sci USA* 108: 14555–14559.

Owen-Smith NR. 1988. *Megaherbivores.* Cambridge, Cambridge University Press.

Passingham RE. 2012. *The neurobiology of the prefrontal cortex.* Oxford, Oxford University Press.

Pearce E, Stringer C, Dunbar RIM. 2013. New insights into differences in brain organization between Neanderthals and anatomically modern humans. *Proc Roy Soc B Biol Sci* 280: 20130.

Pellerin L, Magistretti PJ. 2004. Neuroenergetics: Calling upon astrocytes to satisfy hungry neurons. *J Neuroimaging* 10: 53–62.

Pepperberg IM. 1999. *The Alex studies: Cognitive and communicative abilities of grey parrots*. Boston, Harvard University Press.

Plotnik JM, Lair R, Suphachoksahakun W, de Waal FBM. 2011. Elephants know when they need a helping trunk in a cooperative task. *Proc Natl Acad Sci USA* 108: 5116–5121.

Porter RK, Brand MD. 1995a. Causes of differences in respiration rate of hepatocytes from mammals of different body mass. *Am J Physiol* 269: R1213–R1224.

Porter RK, Brand MD. 1995b. Cellular oxygen consumption depends on body mass. *Am J Physiol* 269: R226–R228.

Prabhakar S, Visell A, Akiyama JA, Shoukry MP, Lewis KD, Holt A, et al. 2008. Human-specific gain of function in a developmental enhancer. *Science* 321: 1346–1350.

Prior H, Schwarz A, Güntürkün O. 2008. Mirror-induced behaviour in the magpie (*Pica pica*): Evidence of self-recognition. *PLoS Biol* 6: e202.

Prüfer K, Racimo F, Patterson N, Jay F, Sankararaman S, Sawyer S, et al. 2014. The complete genome sequence of a Neanderthal from the Altai mountains. *Nature* 505: 43–49.

Raji CA, Ho AJ, Parikshak NN, Becker JT, Lopez OL, Kuller LH, Hua X, Leow AD, Toga AW, Thompson PM. 2010. Brain structure and obesity. *Hum Brain Mapp* 31: 353–364.

Ramnani N. 2006. The primate cortico-cerebellar system: Anatomy and function. *Nat Rev Neurosci* 7: 511–522.

Reiss D, Marino L. 2001. Mirror self-recognition in the bottlenose dolphin: A case of cognitive convergence. *Proc Natl Acad Sci USA* 98: 5937–5942.

Rilling JK, Insel TR. 1999. The primate neocortex in comparative perspective using magnetic resonance imaging. *J Hum Evol* 37: 191–223.

Rolfe DFS, Brown GC. 1997. Cellular energy utilization and molecular origin of standard metabolic rate in mammals. *Physiol Rev* 77: 731–758.

Roth G, Dicke U. 2005. Evolution of the brain and intelligence. *Trends Cogn Sci* 9: 250–257.

Rowe TB, Macrini TE, Luo ZX. 2011. Fossil evidence on origin of the mammalian brain. *Science* 332: 955–957.

Sagan C. 1977. *The dragons of Eden*. Random House, New York.

Sallet J, Mars RB, Noonan MP, Neubert FX, Jbabdi S, O'Reilly JX, et al. 2013. The organization of dorsal frontal cortex in humans and macaques. *J Neurosci* 33: 12255–12274.

Sarko DK, Catania KC, Leitch DB, Kaas JH, Herculano-Houzel S. 2009. Cellular scaling rules of insectivore brains. *Front Neuroanat* 3: 8.

Schmitz C, Hof PR. 2000. Recommendations for straightforward and rigorous methods of counting neurons based on a computer simulation approach. *J Chem Neuroanat* 20: 93–114.

Schoenemann PT, Sheehan MJ, Glotzer LD. 2005. Prefrontal white matter volume is disproportionately larger in humans than in other primates. *Nat Neurosci* 8: 242–252.

Schüz A, Demianenko GP. 1995. Constancy and variability in cortical structure. A study on synapses and dendritic spines in hedgehog and monkey. *J Hirnforsch* 36: 113–122.

Schüz A, Palm G. 1989. Density of neurons and synapses in the cerebral cortex of the mouse. *J Comp Neurol* 286: 442–455.

Seehausen O. 2002. Patterns in fish radiation are compatible with Pleistocene dessication of Lake Victoria and 14600 year history for its cichlid species flock. *Proc Royal Soc Biol B* 269: 1490–1491.

Semendeferi K, Lu A, Schenker N, Damasio H. 2002. Humans and great apes share a large frontal cortex. *Nat Neurosci* 5: 272–276.

Shanahan M, Bingman VP, Shimizu T, Güntürkün O. 2013. Large-scale network organization in the avian forebrain: A connectivity matrix and theoretical analysis. *Front Comput Neurosci* 7: 89.

Sherwood CC. 2005. Comparative anatomy of the facial motor nucleus in mammals, with an analysis of neuron numbers in primates. *Anat Rec* 287: 1067–1079.

Shoham S, O'Connor DH, Segev R. 2006. How silent is the brain: Is there a "dark matter" problem in neuroscience? *J Comp Physiol A Neuroethol Sens Neural Behav Physiol* 192: 777–784.

Shulman RG, Hyder F, Rothman DL. 2009. Baseline brain energy supports the state of consciousness. *Proc Natl Acad Sci USA* 106: 11096–11101.

Silcox MT, Dalmyn CK, Bloch JI. 2009. Virtual endocast of *Ignacius graybullianys* (Paromomyidae, Primates) and brain evolution in early primates. *Proc Natl Acad Sci USA* 106: 10987–10992.

Smaers JB, Steele J, Case CR, Cowper A, Amunts K, Zilles K. 2011. Primate prefrontal cortex evolution: Human brains are the extreme of a lateralized ape trend. *Brain Behav Evol* 77: 67–78.

Smith AJ, Blumenfeld H, Behar KL, Rothman DL, Shulman RG, Hyder F. 2002. Cerebral energetics and spiking frequency: The neurophysiological basis of fMRI. *Proc Natl Acad Sci USA* 99: 10765–10770.

Snell O. 1891. Die Abhängigkeit des Hirngewichtes von dem Körpergewicht und den geistigen Fähigkeiten. *Arch Psychiatr Nervenkr* 110: 2801–2808.

Sokoloff L. 1960. 1843–1864. The metabolism of the central nervous system in vivo. In *Handbook of physiology*, Section 1, *Neurophysiology*, volume 3. Edited by Field J, Magoun HW, Hall VE Washington, D.C., American Physiological Society, pp. 1843–1864.

Sol D, Duncan RP, Blackburn TM, Cassey P, Lefebvre L. 2005. Big brains, enhanced cognition, and response of birds to novel environments. *Proc Natl Acad Sci USA* 102: 5460–5465.

Somel M, Xiling L, Khaitovich P. 2013. Human brain evolution: Transcripts, metabolites and their regulators. *Nature Rev Nsci* 14: 112–127.

Spocter MA, Hopkins WD, Barks SK, Bianchi S, Hehmeyer AE, Anderson SM, et al. 2012. Neuropil distribution in the cerebral cortex differs between humans and chimpanzees. *J Comp Neurol* 520: 2917–2929.

Standage T. 2009. *An edible history of humanity*. New York, Walker & Company.

Stephan H, Andy OJ. 1969. Quantitative comparative anatomy of primates: An attempt at a phylogenetic interpretation. *Ann NY Acad Sci* 167: 370–386.

Stephan H, Frahm H, Baron G. 1981. New and revised data on volumes of brain structures in insectivores and primates. *Folia Primatol (Basel)* 35: 1–29.

Stolzenburg J-U, Reichenbach A, Neumann M. 1989. Size and density of glial and neuronal cells within the cerebral neocortex of various insectivorian species. *Glia* 2: 78–84.

Susman RL. 1998. Hand function and tool behavior in early hominids. *J Hum Evol* 35: 23–46.

Tanaka H, Landmesser LT. 1986. Cell death of lumbosacral motoneurons in chick, quail, and chick-quail chimera embryos: A test of the quantitative matching hypothesis of neuronal cell death. *J Neurosci* 6: 2889–2899.

Thompson DW. 1917. *On growth and form*. Cambridge, Cambridge University Press.

Tocheri MW, Orr CM, Jacofsky MC, Marzke MW. 2008. The evolutionary history of the hominin hand since the last common ancestor of *Pan* and *Homo*. *J Anat* 212: 544–562.

Tower DB. 1954. Structural and functional organization of mammalian cerebral cortex: The correlation of neurone density with brain size. Cortical neurone density in the fin whale (*Balaenoptera physalus L.*) with a note on the cortical neurone density in the Indian elephant. *J Comp Neurol* 101: 19–52.

Tower DB, Elliott KAC. 1952. Activity of the acetylcholine system in cerebral cortex of various unanesthetized animals. *Am J Physiol* 168: 747–759.

Turrigiano GG. 2008. The self-tuning neuron: Synaptic scaling of excitatory synapses. *Cell* 135: 422–435.

Twain, M. [Clemens, SL.] 1973. *What is man? And other philosophical writings.* Edited by Baender, P. Berkeley, University of California Press.

Uddin M, Wildman DE, Liu G, Grossman LI, Goodman M. 2004. Sister grouping of chimpanzees and humans as revealed by genome-wide phylogenetic analysis of brain gene expression profiles. *Proc Natl Acad Sci USA* 101: 2957–2962.

Ventura-Antunes L, Mota B, Herculano-Houzel S. 2013. Different scaling of white matter volume, cortical connectivity, and gyrification across rodent and primate brains. *Front Neuroanat* 7: 3.

Virchow R. 1856. Zur Pathologie des Schädels und des Gehirn. In *Gesammelte Abhandlungen zur wissenschaftlichen Medicin.* Frankfurt am Main, Von Meidinger & Sohn, pp.883–1014.

von Bonin G. 1937. Brain-weight and body-weight of mammals. *J Comp Neurol* 66: 103–111.

von Haller A. 1762. *Elementa physiologiae corporis humani.* Lausanne: Francisci Grasset.

Wadley L, Hodgkiss T, Grant M. 2009. Implications for complex cognition from the hafting of tools with compound adhesives in the Middle Stone Age, South Africa. *Proc Natl Acad Sci USA* 106: 9590–9594.

Walloe S, Eriksen N, Torben D, Pakkenberg B. 2010. A neurological comparative study of the harp seal (*Pagophilus groenlandicus*) and harbor porpoise (*Phocoena phocoena*) brain. *Anat Rec* 293: 2129–2135.

Warneken F, Rosati AG. 2015. Cognitive capacities for cooking in chimpanzees. *Proc Royal Soc B* 282: 20150229.

Watson C, Provis J, Herculano-Houzel S. 2012. What determines motor neuron number? Slow scaling of facial motor neuron numbers with body mass in marsupials and primates. *Anat Rec* 295: 1683–1691.

Watts DP. 1988. Environmental influences on mountain gorilla time budgets. *Am J Primatol* 15: 195–211.

Weir AA, Chappell J, Kacelnik A. 2002. Shaping of hooks in New Caledonian crows. *Science* 297: 981.

Whiten A, Goodall J, McGrew WC, Nishida T, Reynolds V, Sugiyama Y, et al. 1999. Cultures in chimpanzees. *Nature* 399: 682–685.

Williams C. 2012. Are these the brain cells that give us consciousness? *New Scientist,* July 23: 33–35.

Williams RW, Herrup K. 1988. The control of neuron number. *Annu Rev Neurosci* 11: 425–453.

Wilson NR, Ty MT, Ingber DE, Sur M, Liu G. 2007. Synaptic reorganization in scaled networks of controlled size. *J Neurosci* 27: 13581–13589.

Wise SM. 2003. *Drawing the line: Science and the case for animal rights.* New York, Basic Books.

Wobber V, Hare B, Wrangham R. 2008. Great apes prefer cooked foods. *J Hum Evol* 55: 340–348.

Wong P, Peebles JK, Asplund CL, Collins CE, Herculano-Houzel S, Kaas JH. 2013. Faster scaling of auditory neurons in cortical areas relative to subcortical structures in primate brains. *Brain Behav Evol* 81: 209–218.

Wrangham R. 2009. *Catching fire: How cooking made us human.* New York, Basic Books.

Yaman S, Kilian A, von Fersen L, Güntürkün O. 2012. Evidence for a numerosity category that is based on abstract qualities of "few" vs. "many" in the bottlenose dolphin (*Tursiops truncatus*). *Front Psychol* 3: 473.

Yellen JE, Brooks AS, Cornelissen E, Mehlman MJ, Stewart K. 1995. A middle stone age worked bone industry from Katanda, Upper Semliki Valley, Zaire. *Science* 268: 553–556.

Young NA, Flaherty DK, Airey DC, Varlan P, Aworunse F, Kaas JH, Collins CE. 2012. Use of flow cytometry for high-throughput cell population estimates in brain tissue. *Front Neuroanat* 6: 27.

Zhao W-Q, De Felice FG, Fernandez S, Chen H, Lambert MP, Quon MJ, et al. 2008. Amyloid beta oligomers induce impairment of neuronal insulin receptors. *FASEB J* 22: 246–260.

Zimmer C. 2009. The dark matter of the human brain. *Discover,* September.

Index